Modelling of the Impact Response of Fibre-Reinforced Composites

Y. LI, C. RUIZ, J. HARDING

University of Oxford
Department of Engineering Science
Parks Road
Oxford, United Kingdom

CRC Press
Taylor & Francis Group
Boca Raton London New York

CRC Press is an imprint of the
Taylor & Francis Group, an **informa** business

First published by Technomic publishing company, Inc.

Published 2019 by CRC Press
Taylor & Francis Group
6000 Broken Sound Parkway NW, Suite 300
Boca Raton, FL 33487-2742

© 2019 by Taylor & Francis Group, LLC
CRC Press is an imprint of Taylor & Francis Group, an Informa business

First issued in paperback 2019

No claim to original U.S. Government works

ISBN 13: 978-0-367-45069-4 (pbk)
ISBN 13: 978-0-87762-820-0 (hbk)

Visit the Taylor & Francis Web site at
http://www.taylorandfrancis.com

and the CRC Press Web site at
http://www.crcpress.com

CONTENTS

CHAPTER 1

Modelling of the Impact Response of Fibre-Reinforced Composites

Y. Li, C. Ruiz and J. Harding

ABSTRACT: The report describes progress made during the second year of a three-year programme with the above title. On the experimental side an extended series of tensile impact tests on hybrid carbon/glass and carbon/Kevlar laminates has been continued and results obtained are reported. In addition a testing technique has been developed for determining the effect of strain rate on the interlaminar shear strength in composite materials and preliminary results obtained for a woven carbon/epoxy laminate at a quasi-static and an impact rate of loading. A very marked effect of strain rate on both the critical shear strength and the critical shear strain at failure is observed.

On the theoretical side the stress distribution in a hybrid carbon/glass tensile specimen has been determined using the ABAQUS finite element package. This analysis confirms that under tensile loading the first step in the failure process is likely to be the tensile fracture of a carbon fibre tow aligned in the loading direction. By modelling such a failure in terms of a reduced stiffness in the loading direction, the ABAQUS package was used to determine the stress field around the failed tow. Large shear and normal stresses on the interlaminar plane close to the point of initial failure were observed, suggesting that propagation of fracture might be controlled by delamination or by deplying processes. Limited delamination was found, in certain cases, to reduce the magnitude of these local interlaminar shear stresses. A fuller study of the effect of the hybrid lay-up and of the position within the specimen of the initial point of failure on the stress distribution around the failed tow is given in a later chapter.

A previous attempt at using laminate theory to predict the quasi-static hybrid tensile strength has now been extended to include a description of the impact tensile behaviour of the same laminates. Although failure strengths are reasonably well predicted, failure strains are significantly overestimated. A refinement of the analysis, assuming a range of possible tow failure strengths leading to successive single tow failures at increasing loads and taking into account the local stress concentrations around failed tows, is discussed.

Finally, a finite element analysis of the stress distribution around the specimen/ loading-bar interface shows the present rectangular ended design of loading-bar to be as good as any of the more refined geometries studied.

1. INTRODUCTION

Progress on both the experimental and the theoretical aspects of the programme during the first year of the present three-year grant were described in the report issued in May 1988.

The original experimental programme, as proposed under the three Grants preceding the present one, called for tensile stress-strain curves to be determined at quasi-static and impact rates of loading on specimens cut from woven reinforced laminates with all-glass, all-carbon and three different hybrid lay-ups. A similar series of tests was also proposed for all-Kevlar reinforced laminates and for three different carbon/Kevlar hybrid lay-ups. Since a plain weave reinforcement was used it had been assumed that the mechanical properties were the same in both warp and weft directions. Initially the impact tests had been performed for loading in the weft direction while, at a later stage, the quasi-static tests were performed for loading in both the warp and the weft directions and some differences between the two directions observed. During the first year of the present Grant a start had been made on a new set of tensile impact tests in which both carbon/glass and carbon/Kevlar reinforced specimens would be loaded in both the warp and the weft directions and also in a direction in the plane of reinforcement and at 45° to the warp and weft directions. The progress on this part of the experimental work is described in Section 4.1 of the present report.

Also during the first year of the present Grant a start was made on the development of numerical and analytical techniques for the modelling of the experimentally observed impact response of the hybrid laminates. The stress distribution in the standard design of tensile specimen was determined, using the PAFEC finite element package, for both an all-glass and a hybrid carbon/glass lay-up. In each case the tensile stresses in the specimen gauge region were significantly higher than the tensile or shear stresses elsewhere in the specimen, leading to the conclusions that the specimen design is satisfactory and that, as normally obtained in practice, the first step in the failure process will be the tensile fracture of an axially-aligned tow of reinforcing fibres. Subsequently a similar result was obtained, for an all-glass lay-up, using the ABAQUS finite element package. With this package triangular elements are available, allowing a more accurate modelling of the waisted geometry of the test specimen. While this result confirmed the conclusions from the PAFEC analysis regarding the specimen design and the nature of the initial failure, a question remained concerning the magnitude of the stress concentrations in hybrid specimens arising from free edge effects at points on the interface between the carbon- and the glass-reinforced plies where they intersect the tapered surface in the waisted region of the specimen. The results of the ABAQUS analysis of a specimen with a hybrid lay-up are presented in Section 3.1 of the present report.

Stress concentrations also arise, due to their different elastic properties, at the interface between the specimen and the loading bars. In order to minimise these stress concentrations and optimise the load transfer between the specimen and the loading bars, the ABAQUS finite element package has also been used to study the effect of changes in the loading bar geometry on the stress distribution in this region. The results of this analysis are described in Section 3.2 of the present report.

In the light of the expectation that the first step in the failure process is likely to be a tensile failure in one of the axially-aligned carbon or glass re-inforcing tows, the PAFEC package had been used in a first attempt at deter-mining the stress distribution around such a failed region. Failure was modelled as a reduction in stiffness in the loading direction over a region (or "link") with the dimensions shown in Figure 1. Large shear and normal stresses on the interlaminar plane close to the failed link are observed, sug-gesting that propagation of failure may be controlled by delamination or deplying mechanisms. This analysis has now been confirmed by means of the ABAQUS package, which allows the use of interfacial elements, and ex-tended to include the stress distribution around a failed link in various hybrid lay-ups for initial failure in either a carbon- or a glass-reinforced ply. The results of this part of the work are briefly outlined in Section 3.3 of the present report and are described in detail in a separate report, a copy of which is attached as Chapter 2.

As a result of these analytical studies, which show very high normal and shear stresses on the interlaminar planes close to a failed link, the impor-tance of the critical stress levels for deplying and for interlaminar shear fail-ure in controlling the next stage in the failure process has become apparent. Experimental measurements are required, therefore, of the effect of strain rate on both the interlaminar shear strength and the strength normal to the interlaminar plane. Section 2 of the present report describes the develop-ment of a test technique for determining the effect of strain rate on the in-terlaminar shear strength of woven reinforced carbon/epoxy laminates. The technique was also described in a paper presented at the *4th Oxford Interna-tional Conference on Mechanical Properties of Materials at High Rates of Strain* in March 1989 and a reprint of that paper is attached as Chapter 3.

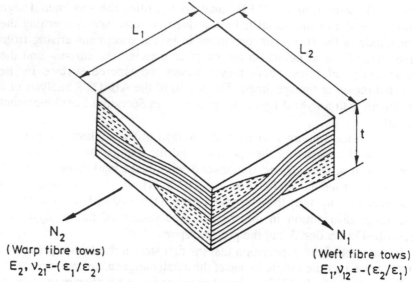

FIGURE 1. Elemental segment (link) in a single fibre-reinforced ply.

In earlier attempts at modelling the tensile stress-strain response of the various hybrid lay-ups a laminate theory approach had been used to predict the behaviour of the hybrid laminates from the experimentally determined response of the individual reinforcing plies. This approach was successful in predicting the tensile modulus at both quasi-static and impact rates of loading and in predicting the quasi-static tensile strength, although the failure strain was considerably overestimated. The analysis has now been extended to the prediction of the hybrid failure strength under impact loading, as described in Section 3.4 of the present report and, in more detail, in a separate report attached as Chapter 4.

2. EXPERIMENTAL TECHNIQUES

2.1 Introduction

Several attempts have been made to develop a test for determining the interlaminar shear strength of composite materials at high rates of strain. Of those based on the Hopkinson-bar technique one of the first was that described by Parry and Harding [1] in which a thin-walled tubular specimen is sandwiched between the input and output bars of a torsional split Hopkinson's bar apparatus. Although this is not an ideal design of specimen for composite materials, when laminated composites are being studied and provided the axis of the tubular specimen is perpendicular to the plane of reinforcement, it should allow some estimate of the interlaminar shear strength at nominal shear strain rates of about 1000/s. In such tests both plain-weave and cross-ply reinforced glass/epoxy laminates, over the limited range of strain rate achievable in the torsional SHPB, showed a marked increase in the apparent interlaminar shear strength with strain rate, see Figure 2. In neither case, however, was failure entirely by simple shear on the interlaminar plane. The woven material showed a spiral failure plane cutting across several reinforcing plies while the cross-ply material failed on several neighbouring interlaminar planes joined by steps inclined at 45° to this plane, see Figure 3.

A second attempt to determine the high strain rate shear strength of composite materials which also uses the torsional SHPB has been made by Chiem and Liu [2]. Their specimen consisted of two strips of square cross section which connected the two loading bars at diametrically opposite points, leaving a cuboid gauge region between the two bars as shown schematically in Figure 4. The results obtained, which are clearly an average response for the two specimens in each test, again showed a marked rate de-

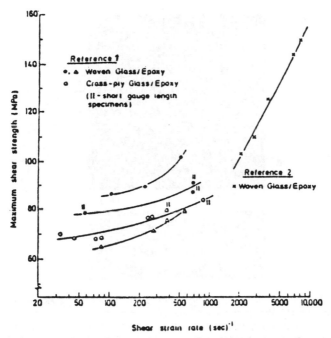

FIGURE 2. Effect of strain rate on maximum shear strength (Ref. [1]).

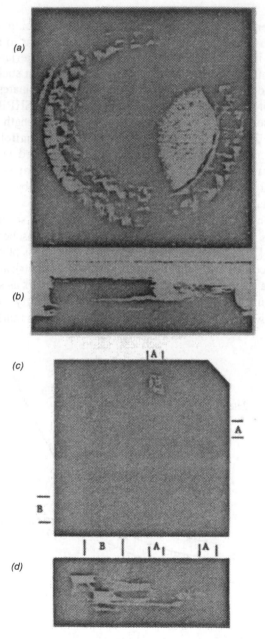

FIGURE 3. Fracture appearance of woven and cross-ply glass/epoxy specimens under torsional impact (Ref. [1]). Woven specimen viewed (a) perpendicular to and (b) parallel to interlaminar plane, showing spiral fracture surface crossing several woven reinforcing plies (×5). Cross-ply specimen viewed (c) perpendicular to and (d) parallel to interlaminar plane: (A) steps at 45° between adjacent interlaminar planes and (B) steps at 45° across 3 adjacent laminae.

Specimens

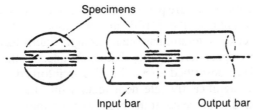

Input bar   Output bar

FIGURE 4. Using torsional Hopkinson-Bar and cuboid specimens (Ref. [2]).

pendence for the shear strength of woven glass/epoxy composite, as also shown in Figure 2. However, the extremely high failure strains reported, up to 80%, raise doubts about the accuracy of the Hopkinson-bar technique for determining strains with this particular specimen configuration and there is also concern regarding the true state of stress within the specimen.

An alternative arrangement of the SHPB, which allows very high shear strain rates to be achieved in metallic materials, uses the double notch shear specimen [3]. A modified version of this technique has been developed by Werner and Dharan [4] to test plain weave carbon/epoxy laminates in both interlaminar and transverse shear at nominal strain rates from 6000 to 18,000/s. The test arrangement is shown schematically in Figure 5. Since un-notched specimens are used the strain rates are, presumably, related to a nominal shear zone width, probably the clearance between the input and

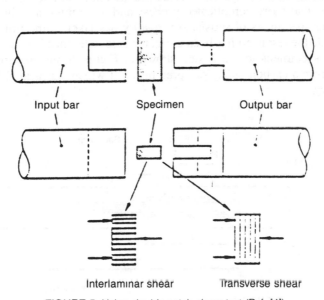

Input bar   Specimen   Output bar

Interlaminar shear   Transverse shear

FIGURE 5. Using double-notch shear test (Ref. [4]).

output bars. Strains, which are particularly difficult to measure with any accuracy in this version of the SHPB, are not quoted. Despite some experimental scatter the interlaminar shear strength determined in these tests for carbon/epoxy specimens, unlike that determined in the other two investigations briefly described above for glass-reinforced specimens, was not found to be significantly rate dependent while the transverse shear strength was found to decrease at the highest strain rates.

2.2 Interlaminar Shear Test for Hybrid Specimens

In the present investigation it was required that the shear strength be determined, at both quasi-static and impact rates of loading, on interlaminar planes between (1) two carbon-reinforced plies, (2) two glass-reinforced plies and (3) one carbon- and one glass-reinforced ply. This required a specimen with a pre-chosen failure plane. Several designs of specimen were considered, that finally chosen being shown in Figure 6. Specimens to this design could not be cut from the same pre-cured woven reinforced laminates as were used for the tension and compression specimens in the earlier stages of the present investigation. For comparable results to be obtained, therefore, the specimens have to be fabricated using the same wet hand lay-up procedure and the same woven glass and carbon mats and epoxy resin as was previously used by Fothergill and Harvey in the manufacture of the original laminates. Compared with the production of simple laminates this is a somewhat more complicated process and developing a technique for producing specimens of satisfactory quality has taken some time. In the meanwhile the test itself has been evaluated (see Reference [5] and Chapter 3) using specimens to the design shown in Figure 6 but prepared more simply from a carbon/epoxy pre-preg material which was available in the laboratory.

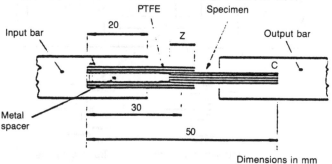

FIGURE 6. Using tensile Hopkinson-Bar (Ref. [5]).

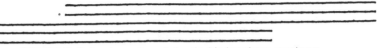

FIGURE 7. Finite element mesh for double lap shear specimen.

2.3 Specimen Preparation

Specimens were prepared from a 5-end satin weave carbon/epoxy pre-preg, woven from 3000 filament fibre tows with, respectively, 70 and 72 yarns per 10 cm in the warp and weft directions and having a dry weight of 285 g/m². The pre-preg was manufactured by Hexcel and Genin using a type ES.36 self-adhesive epoxy resin to give a fibre weight fraction of 52% and an uncured pre-preg weight of 548 g/m². Eight layers of pre-preg were laid up, using metal and PTFE spacers as shown in Figure 6, and then covered by a sealing sheet and evacuated to 28 inches of mercury. This was then placed inside a closed container at an air pressure of 90 psi and the container placed inside a small oven. The temperature was raised over a period of an hour to 125°C, held for 2 hours and then allowed to cool to room temperature.

2.4 Finite Element Analysis of Shear Specimen

The finite element mesh used in a two-dimensional stress analysis of the shear specimen of Figure 6 is shown in Figure 7. Since the specimen is symmetrical about the centre line only half is modelled. There are 120 isoparametrical elements, each with eight nodes. Several different hybrid lay-ups have been analysed only one of which, the all-carbon lay-up shown in Figure 8, being relevant to the present tests. The elastic properties not having been determined for the satin weave carbon pre-preg laminates those previously determined for the plain weave all-carbon laminates were used in the finite element analysis so the results obtained only give a general indication of the stress and strain variation within the specimen.

The results of this analysis are presented in Figure 8. This shows the variation of longitudinal strain, normal strain and shear strain on the interlaminar plane over the 7 mm shear zone length (region X-X) and for a further 3.5 mm to each side. Very large shear strain (and hence stress) concentrations are apparent at each end of the interlaminar plane, points "X" in Figure 8. This problem arises, however, in all designs of interlaminar shear specimen so far considered. In determining the interlaminar shear strength from the measured externally applied load at failure it is necessary to assume both that this load is shared equally between the two failure surfaces

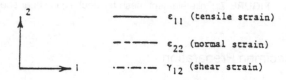

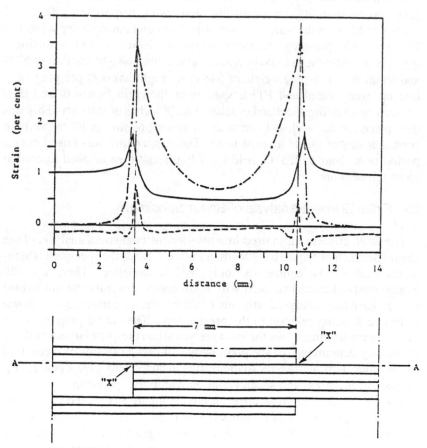

FIGURE 8. Strains on failure surface A-A of interlaminar shear specimen.

and that the required shear strength is related to the average, rather than the peak, value of stress on the failure plane. Modifications to the design of the interlaminar shear specimen in which this large variation in shear stress (and strain) will be reduced are still being considered. Meanwhile results obtained with the specimen described here are presented in Section 4.2.

3. RESULTS OF THEORETICAL ANALYSES

3.1 Stress Distribution in a Hybrid Carbon/Glass Specimen Type 2b

The finite element mesh for the analysis of the type 2b hybrid specimen, including the loading bar to which it is attached, is shown in Figure 9. The hybrid lay-up consists of alternating carbon and glass reinforced plies with a total of two glass and three carbon plies in the cross section of the central parallel region. The nominal thickness of the carbon and the glass plies is 0.28 and 0.08 mm, respectively. Different elastic properties were assumed for the two-ply types in the longitudinal and transverse directions, directions 1 and 2 in Figure 9. As described in an earlier report, Reference [6], the elastic properties in the 1-3 plane were the average of a number of measurements made in quasi-static tests on glass or carbon reinforced specimens, giving, for the all-glass specimens

$$E_1 = E_3 = 16.9 \text{ GPa}, G_{13} = 1.61 \text{ GPa and } \nu_{13} = \nu_{31} = 0.16$$

while in the thickness direction, i.e., the 1-2 plane, arbitrarily chosen (but reasonable) values of

$$E_2 = 12 \text{ GPa}, G_{12} = G_{23} \text{ and } \nu_{21} = 0.15, \text{ giving } \nu_{12} = 0.11$$

were assumed. Corresponding values for the carbon plies were taken to be

$$E_1 = E_3 = 45.3 \text{ GPa}, G_{13} = 2.73 \text{ GPa, and } \nu_{13} = \nu_{31} = 0.13$$

and

$$E_2 = 12 \text{ GPa}, G_{23} = G_{32} \text{ and } \nu_{21} = 0.15, \text{ giving } \nu_{12} = 0.04$$

The resulting calculation showed the longitudinal tensile stresses, i.e., in direction 1, in the parallel gauge section of the specimen to exceed the normal stresses (direction 2) and the interfacial shear stresses (those on the 1-3

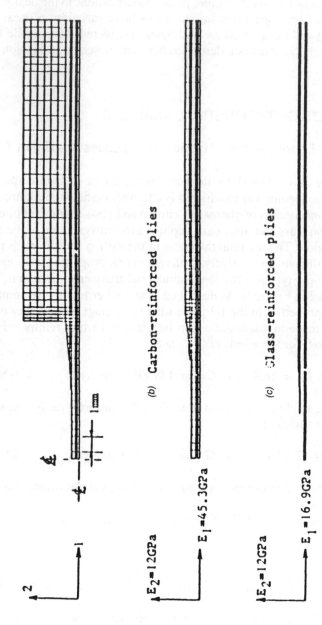

(a) Specimen and loading-bar

(b) Carbon-reinforced plies

$E_1 = 45.3$ GPa

$E_2 = 12$ GPa

(c) Glass-reinforced plies

$E_1 = 16.9$ GPa

$E_2 = 12$ GPa

FIGURE 9. Finite element mesh for hybrid carbon/glass tensile specimen: (a) specimen and loading-bar; (b) carbon-reinforced plies; (c) glass-reinforced plies.

plane) by some five orders of magnitude, confirming that this part of the specimen is effectively in a state of uniaxial tension.

The corresponding stress states in the tapered region of the specimen, i.e., along the interface between the carbon- and the glass-reinforced plies where they intersect the free surface, positions A-B-C and D-E-F in Figure 10, are shown in Figure 11. Stresses are determined at the Gauss integration points close to the nodes. Along the interface A-B-C the normal stresses are three or more orders of magnitude less and the shear stresses two orders of magnitude less than the local longitudinal stresses while along the interface D-E-F this difference is reduced to about two orders of magnitude for the normal stresses and one order of magnitude for the shear stresses. Close to point D, however, discrepancies are apparent in the values of the normal stress and the shear stress when determined on opposite sides of the interface, suggesting that for these stresses the calculation is not very accurate. In contrast, at both interfaces a good agreement is obtained between the values of longitudinal stress measured on either side of the interface when the difference in elastic moduli in direction 1 for the two types of reinforcing ply is taken into account.

These results confirm, for the hybrid lay-up, the previous conclusion for the all-glass lay-up that the design of specimen is satisfactory and that an initial tensile failure determined by the magnitude of the ruling longitudinal stresses in the parallel gauge section, as observed in practice, is the expected failure mode. In addition, for the hybrid lay-up, where there are distinct differences in the elastic properties of the two types of reinforcing ply, it is now shown that stress concentrations at the free surface remain relatively small.

3.2 Stress Distribution at Specimen/Holding Bar Interface

As shown in Figure 12, along the interface between the holding bar and the outer carbon ply of the specimen, nodes G-H-I-J in Figure 10, both the shear stresses and the normal stresses are of the same order as the longitudinal stresses. However, close to the discontinuity at G large discrepancies are observed between the stress values, including the longitudinal stresses after allowing for the modular ratio, as determined on either side of the interface, suggesting that here the finite element analysis is only very approximate. Although the calculated values of shear and normal stress decline very rapidly within a few elements from G, the discrepancies across the interface remain.

By modifying the geometry of the holding bar in the region close to the node at G an attempt has been made to minimise these very high values of shear stress and normal stress and to improve the agreement between their

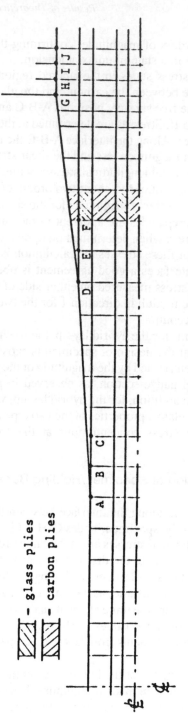

FIGURE 10. Waisted region of hybrid tensile specimen.

- glass plies
- carbon plies

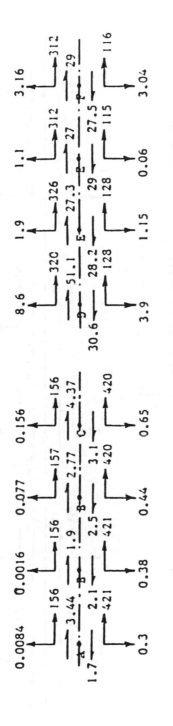

Interface A–B–C

Interface D–E–F

FIGURE 11. Stresses near free surface in waisted region of hybrid tensile specimen.

15

16

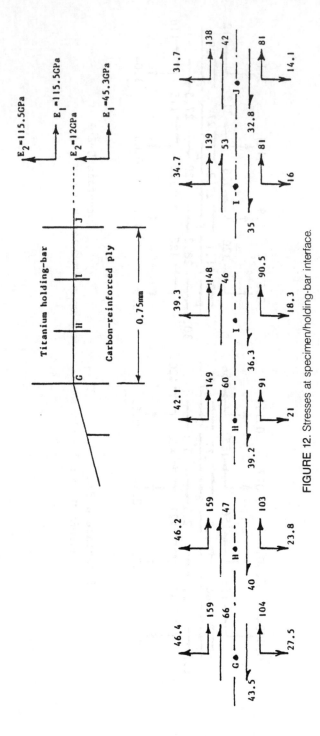

FIGURE 12. Stresses at specimen/holding-bar interface.

calculated values on either side of the interface. Four different designs of loading bar with ends of arbitrarily chosen geometry have been studied. The corresponding finite element meshes, designated WB1, WB2, WB3 and WB4, respectively, are shown in Figure 13. An all-glass specimen with orthotropic elastic properties, as in Figure 9, was assumed since this gave the greatest difference in elastic properties between the specimen and the loading bar. Results are presented in Figures 14 and 15 for the shear stress, S_{12}, and the normal stress, S_2, within the composite along the interface with the holding bar over a distance of 15 mm (50 elements) from the node at G. In each case results for the standard rectangular design of holding bar, designated RE, are compared with those for the four modified designs. As an indication of the validity of the analysis, the difference between the two values of S_{12}, as determined on either side of the interface, is shown in Figure 16 and the difference between the two values of S_2 similarly in Figure 17.

A very similar shear stress distribution is seen in Figure 14 for all loading bar geometries at distances greater than about 1.3 mm from G. Closer to G the rectangular bar shows the highest values of shear stress but the effect is not particularly marked and the trend of an increasing shear stress in the region of G is shown by all five geometries. Also, the discrepancies in the measured values of the shear stress on either side of the interface only become significant within about 1 mm of G (see Figure 16) indicating that the results in this region are likely to be unreliable. None of the modified loading bar geometries, therefore, can be considered to have led to a significant reduction in the shear stress concentration at G.

As may be seen from Figure 15, the normal stress, S_2, is also hardly affected by the loading bar geometry at distances greater than about 3 mm from G. For positions closer to G the rectangular bar initially shows the highest normal stress levels but within 0.5 mm of G these become much the smallest and fall almost to zero. It should be noted, however, that for all loading bar geometries, while the calculated values of normal stress on the specimen side of the interface are relatively low those determined in the loading bar are some 5 × greater, leading to very large discrepancies across the interface, as shown in Figure 17. Since this discrepancy is least for the rectangular design of loading bar there seems to be no reason to change from this design in an attempt to minimise the normal stresses. While a greater improvement might be obtained if the modified geometry were to extend further than the maximum of 4.2 mm, in design WB3, there would be a corresponding loss of accuracy in the elastic wave analysis, which assumes bars of constant cross section. Again, therefore, it must be concluded that the inaccuracies in the analysis close to the node at G are such as to mask any improvement in the stress distribution in this region which might arise from the modifications to the loading bar geometry.

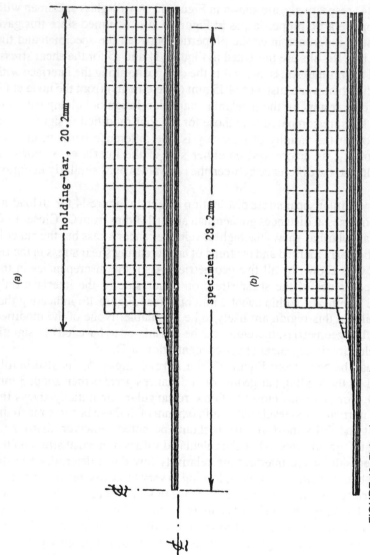

FIGURE 13. Finite element meshes for analysis of shaped holding-bars: (a) first modification (WB1); (b) second modification (WB2).

18

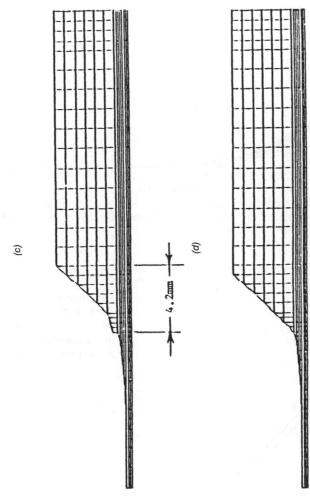

(c)

(d)

4.2mm

FIGURE 13 (continued). Finite element meshes for analysis of shaped holding-bars: (c) third modification (WB3); (d) fourth modification (WB4).

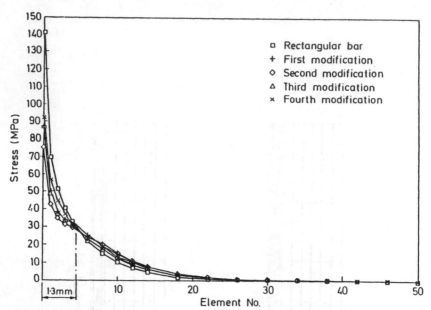

FIGURE 14. Effect of holding-bar design on shear stress variation along specimen/holding bar interface (GFRP specimen: shear stress determined on specimen side of interface).

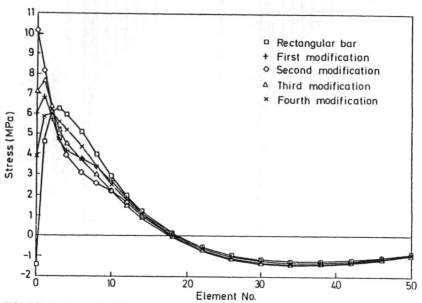

FIGURE 15. Effect of holding-bar design on normal stress variation along specimen/holding bar interface (GFRP specimen: normal stress determined on specimen side of interface).

20

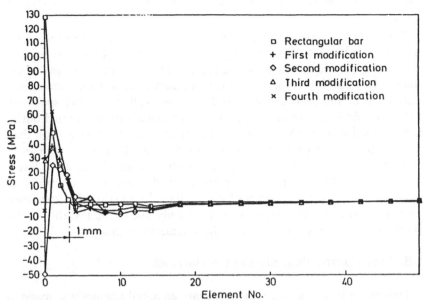

FIGURE 16. Difference between shear stresses determined on either side of specimen/holding-bar interface (GFRP specimen).

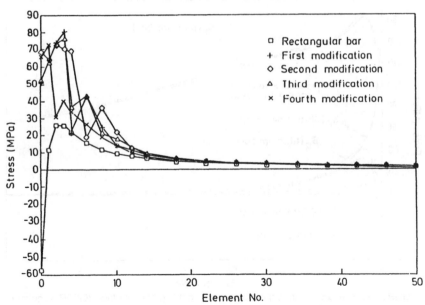

FIGURE 17. Difference between normal stresses determined on either side of specimen/holding-bar interface (GFRP specimen).

21

The normal, shear and longitudinal stress distributions along the specimen/loading-bar interface for the rectangular design of loading bar are shown in Figure 18 for stresses calculated both in the specimen and in the loading bar. To allow comparison the longitudinal stress in the specimen is multiplied by the ratio of the moduli for the loading bar and specimen materials. Very good agreement across the interface is shown throughout by the longitudinal stresses and by the shear stresses at distances greater than 1 mm from G. Although the discrepancy between the normal stresses is much larger and continues over a greater distance from G, the stress levels are low compared with the levels of longitudinal stress. The only problem is the apparently very high level of shear stress close to G on the specimen side of the interface but this is almost certainly due to the inability of the finite element analysis to cope with the singularity at this point.

3.3 Stress Distribution around a Failed Link

During the first year of the present grant an initial attempt was made to determine the stress distribution around a failed link, using the PAFEC finite element package. The analysis showed, not surprisingly, that large

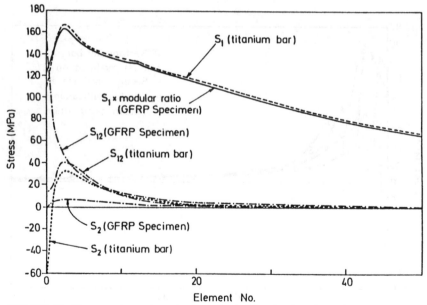

FIGURE 18. Stresses at specimen/rectangular holding-bar interface (GFRP specimen, titanium holding-bar; S_1 = longitudinal stress, S_2 = normal stress, S_{12} = shear stress).

shear stresses were developed on the neighbouring interlaminar planes. It was decided, therefore, to investigate the use of the ABAQUS program and, in particular, the interfacial elements available with this package.

Assuming that the first stage in the fracture process is the tensile failure of one of the carbon fibre tows, as is observed in practice, then to determine whether the next stage is also controlled by the longitudinal stresses or whether a delamination mechanism related to the local shear stresses or a deplying mechanism under the local normal stresses becomes the controlling process it is necessary to determine the redistributed stress system in the vicinity of the first failure. The carbon and the glass-reinforced plies are divided into a number of elements, or links, and first failure is assumed to occur in an arbitrarily chosen link in one of the carbon-reinforced plies.

To model the tensile failure of the given link the longitudinal modulus is reduced by a factor of 0.001. This leads to a singularity at each node of the failed link and to high shear stresses on the interface with the neighbouring plies at points close to these nodes. However, discontinuities in the normal and shear stresses across the interface close to the singularity indicate that here again the analysis has become inaccurate and the stress levels determined cannot be relied upon. This makes difficult the identification of the next stage in the failure process from a comparison of these stresses with the critical stresses for a tensile, a normal or a shear failure at the given strain rate.

In this stress analysis a perfect bond between neighbouring plies, i.e., compatibility of strain between the two types of ply, has been assumed. The corresponding distorted mesh for the failure of a link in the central carbon ply when an overall strain of 2% is applied to the parallel gauge region of the specimen is shown in Figure 19. If the resulting tensile strain concentration in the neighbouring glass- or carbon-reinforced plies is sufficiently high their tensile failure may be the next stage in the failure process. Alternatively the shear strain concentration on the interlaminar plane between the failed ply and the neighbouring glass-reinforced plies could lead to delamination as the next stage in the failure process. Which type of failure follows is determined, therefore, by the ratio of the critical tensile strain to the critical shear strain. Estimates of the former are available from the tensile tests on the non-hybrid carbon- and glass-reinforced laminates while, as mentioned in Section 2.4, methods of determining the critical condition for interlaminar shear failure are still being investigated. However, it should be noted that the marked variation in interlaminar shear strain close to points "X" predicted for the interlaminar shear specimen (see Figure 8) closely mirrors that on the interlaminar planes associated with the failed link (see Figure 19).

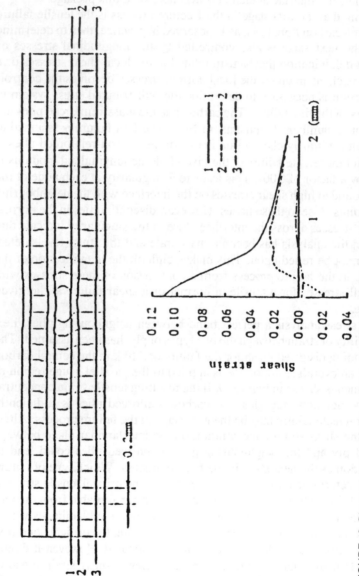

FIGURE 19. Distorted mesh around a failed carbon-reinforced link and variation in shear strain on neighbouring carbon/glass interfaces.

Experimental evidence suggests that, for woven reinforced laminates, a limited delamination follows tensile failure of a given fibre tow, the extent of delamination being greater under impact loading. In the model being considered, if delamination is represented simply as a crack between two frictionless surfaces the shear strain concentration at the singularity is not reduced by the delamination process but merely moves with the delamination crack tip, i.e., the delamination extends catastrophically through to the ends of the specimen. While such behaviour is sometimes observed, particularly in tests on unidirectionally-reinforced specimens, it does not describe the response seen in the present woven hybrid composites. Here, therefore, an arbitrarily chosen friction coefficient of 0.5 is assumed on the delamination crack surfaces between which slip is allowed to occur and the delamination crack is modelled using the ABAQUS interface element for which, since there should be no relative movement until slip occurs, a modulus of 1000 GPa is assumed.

The resulting distorted mesh shapes, for delamination extending over either two or five links on either side of the initially failed link, are shown in Figure 20. On the basis of these analyses delamination does appear to lead to a reduction in the peak shear strain on the interlaminar surface at or near the delamination crack tip, by about 20% for a 2-link delamination and by about 27.5% for a 5-link delamination. This is not a large reduction. Whether it is significant or not may depend on the balance of probability between tensile and shear failure in any given case and on how sensitive shear failure is to the very localised shear concentration close to the link. As shown in Figure 19, even without delamination the shear strain at the interface has fallen by more than an order of magnitude within 5 links, i.e., within 1 mm, of the singularity.

This approach has been extended to investigate the effect of initial failure occurring in a non-central carbon ply or in either the inner or the outer glass-reinforced ply of the stacking sequence shown in Figures 19 and 20. Other stacking sequences and the effects associated with failure in more than one ply have also been studied. The results obtained are fully described in a separate report, included here as Chapter 2. Athough the finite element method has been shown to highlight the features found in the experimental work it is concluded that further studies are needed to characterise the interply shear strength before a full numerical analysis can be developed to model the actual failure process in the tensile specimen. However the importance of the hybrid stacking sequence is clear from the effect it has on the magnitude of the shear strain concentration and on the variation of shear strain on the interlaminar plane for a given order in which tensile fracture and delamination are assumed to occur.

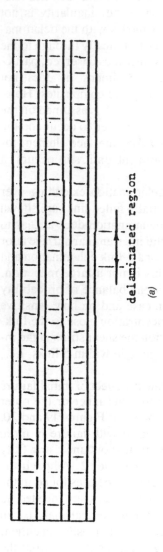

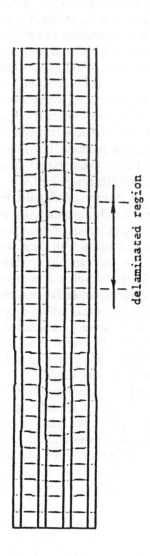

FIGURE 20. Effect of delamination on distorted mesh around a failed carbon-reinforced link: (a) Delamination extending over 2 links (0.4 mm) either side of failed link; (b) delamination extending over 5 links (1.0 mm) either side of failed link.

3.4 Laminate Theory Predictions of Tensile Impact Strength

As has been described in previous reports (References [7,8]) the composite is considered to be made up of a number of layers or plies each containing both reinforcing fibres and the matrix resin. Laminate theory is used to predict, for a given hybrid lay-up, the overall composite properties in terms of the experimentally-determined anisotropic in-plane strengths and stiffnesses of each type of reinforcing ply, in which the properties of the matrix and the given reinforcing fibres are combined. The analysis assumes elastic behaviour in the plies and so can be used to predict the composite elastic moduli. Given a suitable failure criterion and the failure strengths of the reinforcing plies in the principal directions it can also be used to predict "first ply failure", corresponding effectively to the limit of the composite elastic response. This may not, of course, be the same as final overall failure of the composite, in which case some way of accounting for already failed plies in continuing the analysis beyond first ply failure will be required if final failure is also to be predicted.

Using as the failure criterion the two-dimensional form of the Tsai-Wu criterion (Reference [9]),

$$(\sigma_1^2/XX') + (\sigma_2^2/YY') - (\sigma_1\sigma_2)/(XYX'Y')^{0.5} + (\sigma_6^2/SS')$$

$$+ \sigma_1(X' - X)/XX' + \sigma_2(Y' - Y)/YY' + \sigma_6(S' - S)/SS' = 1 \quad (1)$$

where X, X', Y and Y' are the tensile and compressive failure strengths of the individual plies in the principal directions of reinforcement, and S and S' are the corresponding in-plane shear strengths, the first ply to fail is that for which the applied stresses, determined by the application of laminate theory, first satisfy the corresponding form of Equation (1).

For loading in either the warp or the weft direction the in-plane shear strengths, S and S', disappear from Equation (1) so the predicted first ply failure strength depends only on experimentally determined values of the tensile and compressive strengths, X, X', Y and Y', at the appropriate rate of strain. In practice, for this particular loading configuration the values taken for the compressive strengths do not have a significant effect on the predicted first ply failure strength. Following first ply failure the stresses within the composite will be redistributed. The mechanisms which control this redistribution are likely to be complex, involving such processes as matrix cracking, fibre fracture and delamination. A common way of allowing for this is to assume that the "failed" plies can still support a load and to use a reduced stiffness matrix in calculating that proportion of the increased load on the composite which they carry. The loading is then increased until a second critical value is reached at which the stress system on the other set of reinforcing plies satisfies Equation (1) at which point overall failure takes place.

The resulting predicted stress-strain curve is a bilinear approximation to the experimentally determined hybrid mechanical response. In tests at a quasi-static loading rate the agreement between the experimental and the predicted stress-strain curves was not unreasonable, as shown in Figure 21. For all three hybrid lay-ups studied the carbon-reinforced plies failed first, giving the "knee" effect often observed in practice with woven reinforced composites. However the stress level at which first ply failure was predicted lay significantly above the experimentally observed position of the knee. Also, while the predicted failure strengths were closer to the experimental values than the rule of mixtures predictions, the predicted failure strains were all overestimated, particularly at the highest carbon volume fraction.

Inherent in the analysis described above is the assumption of a linear elastic stress-strain response through to failure of the non-hybrid carbon and the non-hybrid glass laminates. While this is not too unreasonable an approximation in the quasi-static tests, at impact rates of strain the stress-strain curve for the non-hybrid glass laminates shows a marked knee effect and a reduced stiffness in the region following this knee, see Figure 22. In an attempt to allow for this (see Reference [8]) first failure in the glass-reinforced plies under impact loading has been related to the stress system associated with the "yield strength", i.e., with the strength parameters X_y

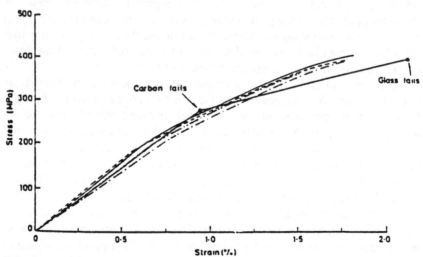

FIGURE 21. Comparison of experimentally measured and laminate-theory predicted quasi-static tensile stress-strain curves for hybrid carbon/glass specimen (Ref. [8]).

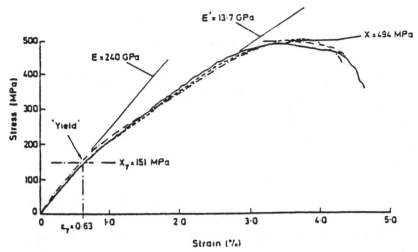

FIGURE 22. Characterisation of tensile impact response of woven all-glass specimen (Ref. [8]).

and Y_y determined at the knee point, see Figure 22. Following this "yield" failure in the glass plies a reduction is assumed in their stiffness matrix in proportion to the reduction of slope of the relevant tensile stress-strain curves, i.e., from 24.0 to 13.7 GPa in Figure 22. A second knee point is then obtained, at first failure in the carbon plies. Overall composite failure corresponds to the subsequent final failure of the glass plies.

The resulting predicted stress-strain response now has three linear regions, as shown in Figure 23 for the type 1 hybrid lay-up impacted in either the warp or the weft direction. In practice, allowing for the reduced stiffness of the glass plies after "yield" appears to have very little effect on the shape of the predicted stress-strain curve up to the point at which the carbon-reinforced plies fail, for loading in either direction. Beyond this point, however, it leads to the prediction of a slightly lower failure strength and a significantly higher failure strain. Since the failure strains are always overestimated, even when the reduced stiffness of the glass-reinforced plies is not taken into account, the comparison between theory and experiment which follows will ignore this effect. This may be supported on the grounds that the simple assumption of an arbitrary reduction in stiffness in all the carbon-reinforced plies at "first ply failure" is too crude a model of the process of damage accumulation to justify the proposed refinement.

Predicted and experimentally determined failure strengths and strains for the three hybrid lay-ups are compared in Table 1. The predicted bilinear stress-strain response for each lay-up is compared with the corresponding

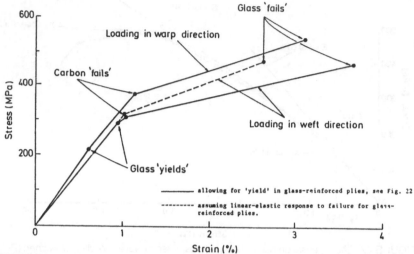

FIGURE 23. Comparison of predicted stress-strain curves for tensile impact tests on type 1 hybrid specimens.

experimentally determined stress-strain curves, as reported earlier (see Reference [7]), in Figures 24, 25 and 26 and a comparison of the effect of the hybrid carbon fraction on the predicted and the measured tensile strengths and failure strains is shown in Figures 27 and 28, respectively.

4. EXPERIMENTAL RESULTS

4.1 Tensile Impact Tests on Carbon/Glass and Carbon/Kevlar Laminates

The laminate theory predictions of hybrid impact tensile strength described in Section 3.4 were compared with experimental data obtained in

TABLE 1. Comparison of Predicted and Experimentally Determined Hybrid Impact Failure Strengths and Strains.

Hybrid Lay-Up	Failure Strength (MPa)		Fracture Strain (%)	
	Experimental	Theoretical	Experimental	Theoretical
Type 1	458	475	1.68	2.61
Type 2a	521	492	1.83	2.63
Type 2b	520	500	1.51	2.64

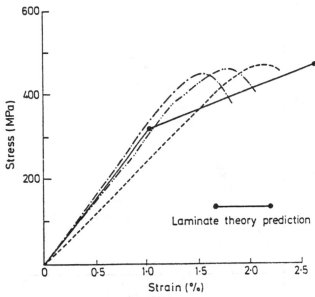

FIGURE 24. Comparison of experimentally-measured and laminate theory predicted stress-strain curves for type 1 hybrid specimens (impact loading in weft direction).

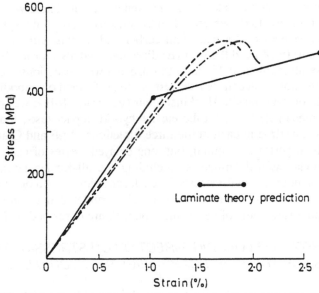

FIGURE 25. Comparison of experimentally-measured and laminate theory predicted stress-strain curves for type 2a hybrid specimens (impact loading in weft direction).

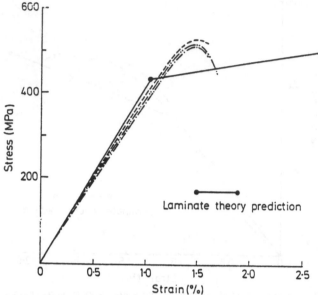

FIGURE 26. Comparison of experimentally-measured and laminate theory predicted stress-strain curves for type 2b hybrid specimens (impact loading in weft direction).

impact tests on specimens loaded in the weft direction. These were the first tensile impact tests to be performed on hybrid specimens. Subsequently impact tests have been performed on all-carbon and all-glass laminates loaded in both the warp (A) and the weft (B) directions and also for loading in the plane of reinforcement at 45° to both warp and weft directions, designated the (C) direction. Since these later tests were performed on a modified version of the original tensile Hopkinson bar (Reference [6]) it was proposed (see Reference [8]) to repeat the earlier hybrid impact tests, loading the specimens this time in each of the three directions, A, B and C and at the same time to perform a similar, but long delayed, series of tests on three carbon/Kevlar hybrid laminates. Including the all-Kevlar laminate and assuming a minimum of three tests for each condition, a total of 63 tests was required. This test programme is now nearly completed and data reduction is well underway. Some of the results obtained are described in Table 2.

4.1.1 EFFECT OF LOADING DIRECTION ON STRESS-STRAIN CURVES FOR THE TYPE 1 CARBON/GLASS HYBRID LAY-UP

Stress-strain curves for three tests on the type 1 carbon/glass laminate impacted in the C direction are shown in Figure 29. Strain gauges attached

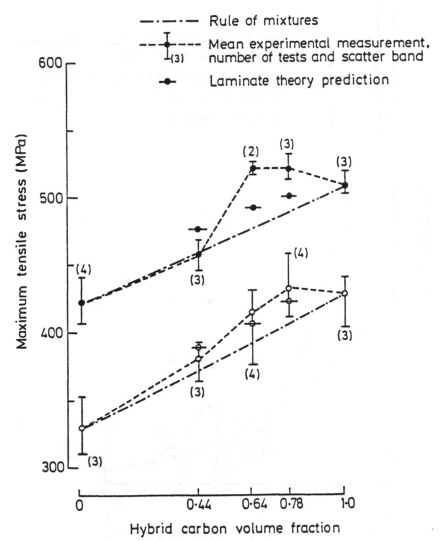

FIGURE 27. Effect of hybrid carbon fraction on maximum tensile stress. Comparison of theory with experiment.
----○---- quasi-static tests, loading in warp direction
---- ●---- impact tests, loaded in weft direction

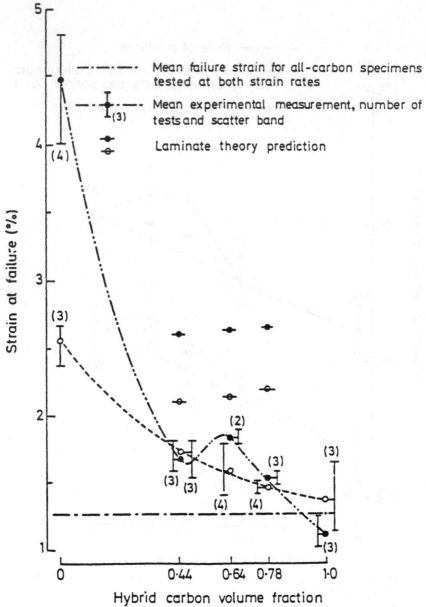

FIGURE 28. Effect of hybrid carbon fraction on strain at failure. Comparison of theory with experiment.
----○---- quasi-static tests, loading in warp direction
----●---- impact tests, loaded in weft direction

TABLE 2. Effect of Loading Direction on the Impact Mechanical Properties of the Type 1 Hybrid Laminate.

Loading Direction	Tensile Modulus (GPa)	Tensile Strength (MPa)	Failure Strain (%)
Warp (A)	44.3 ± 0.00	480.1 ± 4.6	1.34 ± 0.03
Weft (B)	43.8 ± 0.95	436.9 ± 12.4	1.26 ± 0.03
45° (C)	17.4 ± 1.60	246.0 ± 6.8	4.70 ± 0.50

to the specimen were used to check the Hopkinson-bar strain analysis. The degree of scatter is acceptably small. Mean stress-strain curves from these and similar sets of tests on specimens loaded in the A and B directions are compared in Figure 30. Values for the initial elastic modulus, the maximum tensile strength and the strain to failure derived from these tests are tabulated below.

4.1.2 EFFECT OF HYBRID FRACTION ON THE TENSILE IMPACT STRESS-STRAIN CURVES (COMPARISON WITH EARLIER RESULTS)

Mean stress-strain curves for tests on the three carbon/glass laminates loaded in the B direction are compared in Figure 31. Values for the initial elastic modulus, the maximum tensile strength and the strain to failure

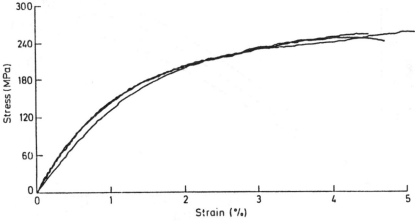

FIGURE 29. Stress-strain curves for tensile impact tests on type 1 carbon/glass hybrid specimens (loading in C direction).

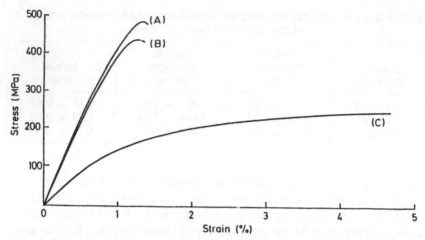

FIGURE 30. Effect of loading direction on mean stress-strain curves for tensile impact tests on type 1 hybrid carbon/glass specimens.

derived from these tests (series 2) are compared in Table 3 with those obtained in earlier impact tests (series 1) on the same laminates. A significant difference between the two sets of results is apparent, the tensile modulus being nearly 50% higher than in the earlier tests and the tensile strength between 4 and 6% lower. A decrease in tensile strength is not surprising in

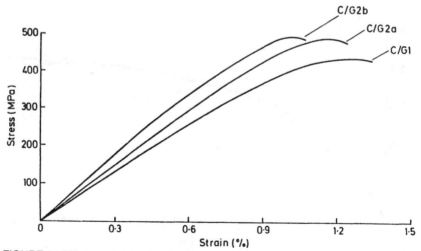

FIGURE 31. Effect of hybrid carbon fraction on mean stress-strain curves for tensile impact tests on specimens loaded in the weft direction.

TABLE 3. Effect of Carbon Hybrid Fraction on the Mechanical Properties for Impact Loading in the Weft Direction.

Hybrid Lay-Up		Tensile Modulus (GPa)	Tensile Strength (MPa)	Failure Strain (%)
Type 1	Series 1	31.6 ± 1.30	458.0 ± 11.0	1.68 ± 0.10
	Series 2	43.8 ± 0.95	436.9 ± 12.4	1.26 ± 0.03
Type 2a	Series 1	33.8 ± 0.60	520.5 ± 1.5	1.83 ± 0.05
	Series 2	48.6 ± 1.15	489.5 ± 5.5	1.18 ± 0.04
Type 2b	Series 1	39.5 ± 0.50	520.0 ± 8.0	1.83 ± 0.05
	Series 2	57.2 ± 0.25	492.7 ± 35.0	1.03 ± 0.03

view of the lapse of time, about 4 years, between the two sets of tests but the increase in modulus is difficult to understand.

The two sets of results are also compared in Figures 32 and 33 which show, respectively, the effect of hybrid fraction on the tensile modulus and the tensile strength. Also included are the results for the all-carbon and all-glass laminates, obtained in tests performed at a time about mid-way between the first and second series of hybrid impact tests.

4.2 Tensile Impact Tests on All-Kevlar Laminates

Stress-strain curves for the first three tensile impact tests on specimens of the all-Kevlar laminate, for loading in the warp (A) direction, are shown in Figure 34. Similar tests have also been performed for loading in the weft (B) and 45° (C) directions and the results will be reported as soon as the data have been analysed. For tests in the A and B directions, 0/90° rosettes were used for the specimen strain gauges, allowing a determination of the transverse strain and hence of Poisson's ratio, ν_{AB} and ν_{BA} in the plane of reinforcement. Table 4 summarises the impact mechanical properties obtained for loading in the A direction.

4.3 Interlaminar Shear Tests on All-Carbon Laminates

Tests have been performed at a quasi-static rate, using a standard Instron loading machine, and an impact rate, using the tensile version of the split Hopkinson-bar apparatus. Interlaminar shear specimens, of the design shown in Figure 6, were fabricated as described previously in Section 2.3. For these initial tests the Instron standard load cell and chart recorder were used to give a load-displacement trace in which the displacement is derived

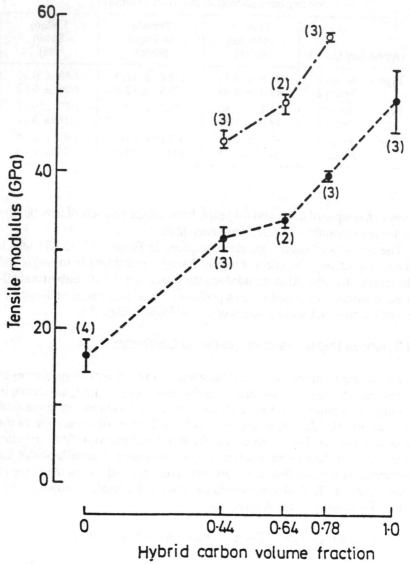

FIGURE 32. Effect of hybrid carbon fraction on tensile modulus for impact loading in the weft direction.

---- ●---- early tests (1984)
----○---- recent tests (1988)

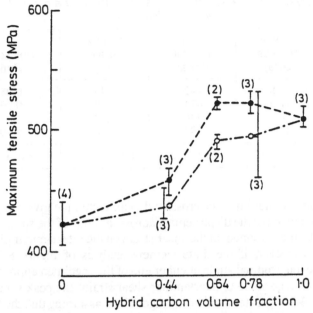

FIGURE 33. Effect of hybrid carbon fraction on maximum tensile stress for impact loading in the weft direction.

---- ● ---- early tests (1984)
---- ○ ---- recent tests (1988)

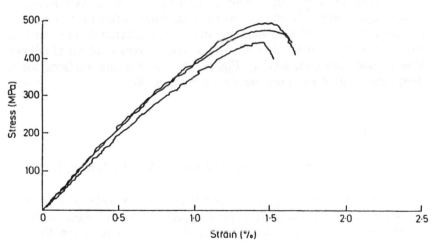

FIGURE 34. Stress-strain curves for tensile impact tests on woven-reinforced Kevlar/epoxy specimens loaded in the warp direction.

TABLE 4. Impact Mechanical Properties of All-Kevlar Laminate Loaded in the Warp Direction.

Test No.	Tensile Modulus (GPa)	Tensile Strength (MPa)	Failure Strain (%)	Poisson's Ratio
139	38.8	442	1.45	0.087
142	41.5	474	1.50	0.095
144	–	436	1.42	–
146	43.1	494	1.50	0.105
	41.3 ± 2.2	461.5 ± 29	1.47 ± 0.04	0.096 ± 0.009

directly from the motion of the crosshead. This technique gives only an approximate estimate of the displacement across the ends of the specimen. No experimental measurement of the shear strain on the interlaminar plane was attempted. However, if the finite element analysis of Figure 8 is to be believed the longitudinal strain at either end of the specimen approximately equals the average value of interlaminar shear strain, the peak shear strain being some 3.5× greater than the average. Thus, assuming that the load applied to the specimen is shared equally between the two interlaminar failure planes, an average interlaminar shear stress-strain curve may be derived from the Instron load-displacement record.

In similar tests under impact loading, strain gauge signals were recorded in the usual way at two gauge stations on the input bar and one on the output bar, a typical set of such records being shown in Figure 35. No strain gauges were attached directly to the specimen. Using the standard Hopkinson-bar wave analysis the displacement across the specimen during the test was determined and, with the assumptions given above, dynamic interlaminar shear stress-strain curves derived. Stress-strain curves so obtained at both loading rates are compared in Figure 36. The fracture surfaces for a dynamically failed specimen are shown in Figure 37.

5. DISCUSSION

5.1 Laminate Theory Predictions of Hybrid Tensile Strength

The laminate theory prediction that failure in the hybrid specimens will initiate in the carbon-reinforced plies appears to be in accord with experimental observations. However, the theoretical assumption that all carbon-reinforced plies fail simultaneously is highly unlikely to be true.

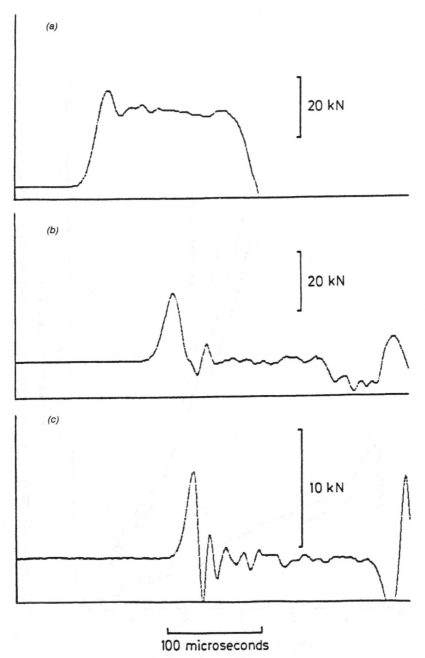

FIGURE 36. Strain gauge signals for impact test on all carbon interlaminar shear specimen: (a) input bar gauge one signal; (b) input bar gauge two signal; (c) output bar gauge signal.

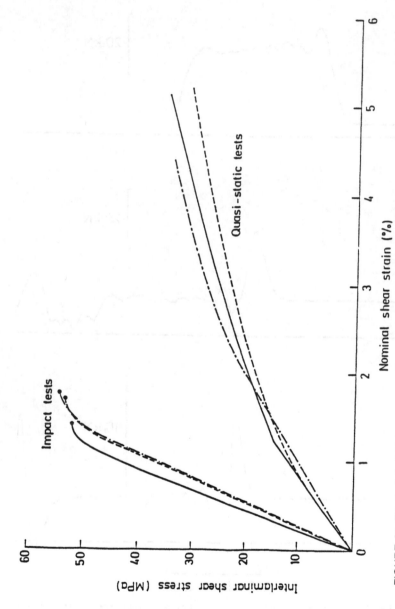

FIGURE 36. Effect of strain rate on stress-strain curves for interlaminar shear tests on woven reinforced carbon/epoxy specimens.

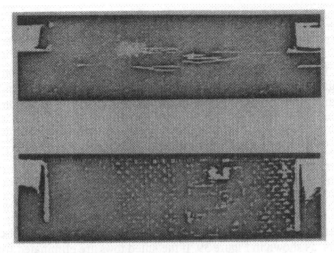

FIGURE 37. Fracture appearance of impacted all-carbon interlaminar shear specimen (×2).

Nor is it likely that their subsequent mechanical response is adequately modelled by a simple reduction in stiffness in the direction of loading. In practice, failure almost certainly initiates at a local weak spot in a single tow of carbon fibres aligned in the loading direction. This will lead to a small reduction in the overall composite stiffness and an increase in the load carried by neighbouring fibre tows. Whether or not they immediately fail under this increased load will depend on how much stronger they are than the fibre tow at the local weak spot. If they do fail immediately the overall stress-strain response is effectively linear-elastic up to failure. In practice a departure from linearity was apparent in all tests, including those on all-carbon lay-ups at impact rates of loading. This implies that there must be a range of failure strengths for the carbon fibre tows from the weakest, at which failure initiates, through to that for those involved in final failure when a particular cross section becomes too highly loaded.

At quasi-static rates the bilinear stress-strain response predicted by laminate theory gives initial failure at a stress level significantly above the first departure from linearity in the actual specimen tests. This is not surprising since the predicted initial failure (knee point) is related to the ultimate tensile strength determined in tests on specimens with an all-carbon lay-up, i.e., it is related to an average strength for the carbon fibre tows, rather than to the strength of the weakest tow. In practice, the strengths range between the first departure from linearity and final failure in the stress-strain curves for the all-carbon tests will give a first estimate of the range of strengths that might be expected between different carbon fibre tows. Bearing in mind that at final failure several carbon fibre tows will have already failed, so that the load carrying cross section of the specimen is reduced, the strength range will in fact be greater than this.

Thus, if we allow one carbon fibre tow to fail in the cross section of the parallel gauge region of an all-carbon specimen, see Figure 38, and if we assume that after failure this tow (a) carries no load and (b) delaminates over the whole of the gauge region, i.e., the region over which the average specimen strain is determined, a reduced longitudinal modulus will result. For impact tests on all-carbon specimens loaded in the weft direction the modulus reduces from 49 GPa in the undamaged state to 47.5 GPa, for one failed carbon fibre tow, and to 46 GPa for two failed tows. The corresponding nominal stresses, i.e., the applied load divided by the original undamaged cross section, may then be obtained from the experimentally determined stress-strain curve for tensile impact tests on the all-carbon specimens, as shown in Figure 39. If the failed tows are eliminated from the cross section, these nominal stresses increase to give average stresses on the reduced cross section of 514 MPa and 544 MPa, respectively. Close to the failed tow, however, finite element studies have confirmed the presence of local stress concentrations. These will give even higher stress levels on the nearby carbon tows. In fact the specific lay-up of Figure 38 has not, as yet, been analysed. However, for a similar lay-up, i.e., that shown in Figure 40, the average value of the longitudinal strain in the fibre tows both imme-

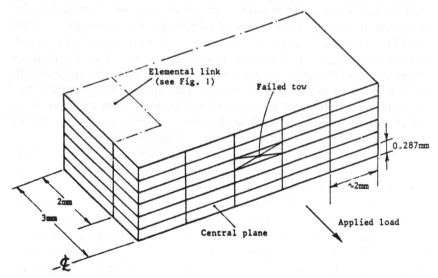

FIGURE 38. Arrangement of elemental links in parallel region of all-carbon tensile specimen (parallel gauge length, 6 mm).

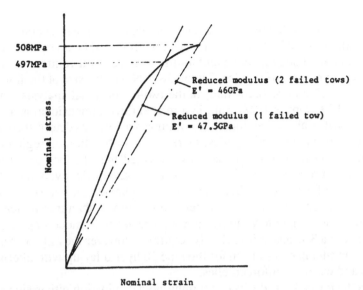

FIGURE 39. Mean experimental stress-strain curve for tensile impact tests on all-carbon specimen (schematic).

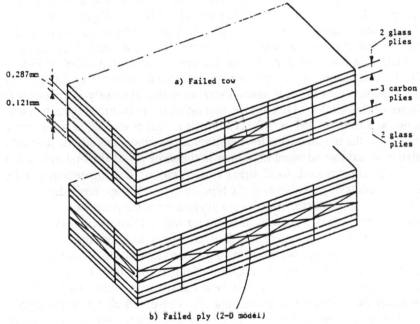

FIGURE 40. Arrangement of elemental links for hybrid lay-up with three adjacent carbon plies.

diately above and below the failed tow exceeded the mean overall applied longitudinal strain by a factor of about 1.8 ×. It should be noted that the size of the characteristic repeating unit for the woven carbon-reinforced ply is 2 mm × 2 mm × 0.287 mm (see Figure 38) whereas the size of the finite element was 0.2 mm × 0.287 mm, in the two-dimensional analysis, and 0.5 mm × 0.595 mm × 0.287 mm, in a related three-dimensional analysis.

The local strain concentration factor quoted above was determined over a single finite element and applies, therefore, to only a limited region of the neighbouring fibre tows. It may also be noted that the 2-D analysis, on which the strain concentration factor of 1.8 is based, effectively models not the failure of a single tow but, as shown in Figure 40(b), that of the whole ply. This gives a much greater reduction in stiffness than that indicated in Figure 39 for single tow failure. For a more accurate modelling of single tow failure a 3-D analysis is clearly required. However the only 3-D analysis so far attempted has been for the type 2b hybrid lay-up with alternating glass and carbon-reinforced plies.

The factor of 1.8 is derived assuming no delamination either side of the failed tow or ply. If limited delamination is allowed, over a region of length 0.4 mm (i.e., two finite elements) either side of the failed element, this factor decreases to about 1.34 × for fibre tows immediately above and below the failed tow and to 1.4 × for the same tows on nearby cross sections (see Chapter 2, Figure 19). In similar calculations [see Figure 20(b)] the delaminated region has been increased to 1 mm (5 finite elements) and a further reduction in the local strain concentration observed. Although this analysis has not been done for the lay-ups of either Figure 38 or Figure 40, by comparison with those which have been studied (see Chapter 2) delamination over the full parallel section is thought likely to reduce this factor to about 1.2 ×. This implies that either the neighbouring carbon tow failed at a stress of 497 × 1.2 MPa or that a weaker tow failed somewhere else while the neighbouring tows continued to support this stress. It should also be noted that adjacent fibre tows in the same ply as the first tow to fail will also experience a local stress concentration. A 3-dimensional finite element analysis (see above) of the type 2b hybrid lay-up shows that, in the absence of delamination, this is also likely to be of the order of 1.34 × and, presumably, will also decrease in a similar way if delamination takes place.

To carry the argument further it is necessary to know the local stress levels near the failed tows following the second tow failure. These are likely to depend critically upon the position of this second tow. The most likely tows to fail second, of course, will be those adjacent to the first failed tow, in either the same or a neighbouring ply. These will also give the largest further increase in local stress. Since none of the finite element calculations of Chapter 2 considers the situation following second tow failure we can only assume, as a first guess, an increase in the strain concentration factor from 1.2 × to 1.4 ×, giving final failure at an approximate local stress of 508 × 1.4 MPa, i.e., an estimated strength range for the carbon fibre tows from 482 to 711 MPa.

The same approach may be used to describe the behaviour observed experimentally in impact tests on hybrid specimens. The cross section of the gauge region of a type 2b hybrid specimen and the mean stress-strain curve for impact loading in the weft direction are shown in Figures 41 and 42 respectively. Again assuming sequential failure of carbon fibre tows and their delamination over the entire gauge length the corresponding reductions in stiffness are calculated and final composite failure is related to the failure of three carbon tows at a nominal failure stress of 520 MPa. The effect of hybridisation may now be clearly seen in the significantly reduced stress (strain) concentrations on the carbon fibre tows closest to each of the failed tows. Although a 2-D finite element analysis has not been performed for the type 2b hybrid lay-up of Figure 41 and the 3-D analysis referred to above considered only shear stress variations in the longitudinal direction, for the very similar lay-up of the type 2a hybrid specimen, a 2-D analysis, assuming initial failure in the central carbon tow (see Figure 43) and Figure 8 of Chapter 2, gave an average local strain concentration factor of only 1.24 × in the neighbouring carbon tows, as compared with 1.80 × in the all-carbon lay-up. This reduces to about 1.16 × after delamination over five elements (Appendix I, Figure 10) and is nearly constant over the delaminated region which corresponds approximately to a half wavelength of the woven reinforcement eithcr side of the failed element. A very similar strain concentration factor of about 1.17 × is obtained if the first element to fail is in an outer carbon-reinforced ply (see Figure 13 of Chapter 2).

If the initial departure from a linear elastic response, i.e., first tow failure, in Figure 42 is observed at a nominal stress of about 300 MPa, the cor-

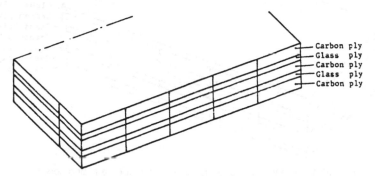

FIGURE 41. Arrangement of elemental links for type 2b hybrid specimen.

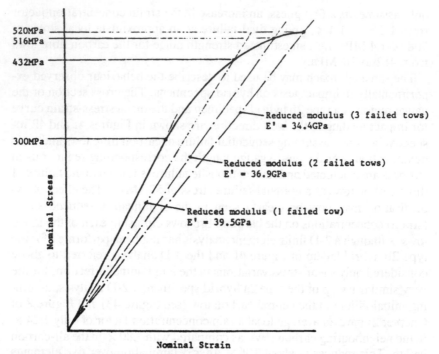

FIGURE 42. Mean experimental stress-strain curve for tensile impact tests on type 2b hybrid specimen (schematic only).

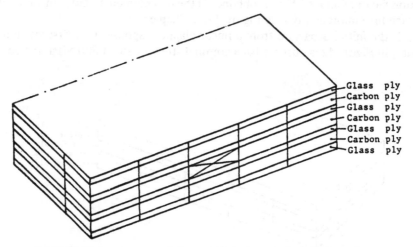

FIGURE 43. Arrangement of elemental links for type 2a hybrid specimen.

responding stress on the carbon-reinforced plies, assuming strain compatibility between carbon and glass plies, is only about 350 MPa, well below the stress of 482 MPa at which the weakest carbon tow apparently failed in the all-carbon tests. If the second carbon tow fails at a nominal stress of 432 MPa (Figure 42) and at a strain (stress) concentration of 1.16 ×, the corresponding local stress on the carbon tow is found to be about 620 MPa, rather closer to the estimated stress of 596 MPa at second tow failure in the all-carbon specimen. It is not possible at this stage to go further with these calculations since finite element analyses are not available from which to estimate the strain concentration factor after the second tow failure. However, should the third carbon tow fail to do so on the same cross section as the first and second, the case illustrated in Figure 16 of Chapter 2 will arise, giving peak tensile strains of between 9 and 10% in the neighbouring glass tows, about twice the critical failure strain in impact tests on all-glass specimens (see Figure 28).

Although this discussion remains somewhat inconclusive it is clear that to refine the laminate theory approach to the prediction of the hybrid tensile impact strength it is necessary to consider a range of values for the critical tensile strength or failure strain for individual carbon fibre tows. The existence of such a range is implied in the observed behaviour of the all-carbon specimens but it is not easy to see how a direct experimental determination of the corresponding statistical distribution could be made. Also, if the hybrid effect is to be investigated it is necessary to determine the effect of the second, high elongation phase, in this case the glass-reinforced plies, on the stress concentrations around failures in the carbon tows. This effect has been illustrated in the above discussion but in the absence of the relevant finite element analyses cannot as yet be used to predict the hybrid tensile impact response.

5.2 Impact Tests on Hybrid Carbon/Glass Specimens

As described in Section 4.1.2 differences have been found between the moduli and tensile strength measurements made in recent impact tests on the three hybrid carbon/glass laminates and those made in an earlier series of tests using the original version of tensile Hopkinson-bar apparatus. The tensile strengths measured in the recent tests are lower than those determined previously by between 4 and 6%. This reduction in strength is almost certainly due to a deterioration in the laminate properties over the four-year period since the earlier tests. However, as may be seen in Figure 33, the variation in tensile strength with hybrid carbon fraction shows the same trends in both series of tests, the type 2a lay-up being stronger and the type 1 lay-up weaker than might have been expected on the basis of the carbon fraction alone. This provides strong evidence that the observed effect is real and not a result of experimental scatter and suggests that alternating plies of the high elongation and the low elongation phase with the high elongation phase on the surface gives the optimum hybrid strength.

The same general trend in the two sets of tests is also observed (see Figure 32) for the hybrid effect on the tensile modulus. Here, however, the recent tests show about a 50% increase over the earlier measurements. In both series of tests the same method was used to determine the modulus, i.e., by means of strain gauges attached directly to the specimen. The only significant difference between the original and the modified versions of Hopkinson bar is in the shape of the input loading wave. In the recent tests this rises very rapidly to a peak stress level early in the tests, as shown, for example, in Figure 35(a). Consequently a higher strain rate is reached earlier in the test, i.e., in the region where the deformation is predominantly elastic. Estimates of the strain rate in the two sets of tests are given in Table 5. If this difference in strain rate is the explanation for the increased elastic moduli determined in the recent tests then they are clearly very strain rate sensitive at rates of this order. It should be noted that, although the moduli for the all-carbon and all-glass specimens were also determined on the modified Hopkinson-bar apparatus, the firing pressure, and hence the strain rate, was lower, explaining their closer agreement with the earlier hybrid results (see Figure 32). Here again the results imply a very strong rate sensitivity under impact loading.

5.3 Interlaminar Shear Strength Tests

Initial results for an all-carbon laminate prepared from woven carbon pre-preg show a marked increase in interlaminar shear strength under impact loading and a significant decrease in the nominal shear strain at failure. It should be noted, however, that the strain is not measured directly and that an average value is assumed although finite element analyses show a variation of some 3.5× in the magnitude of the actual shear strain at different

TABLE 5. Comparison of the Mean Strain Rate in Tests Using Original and Modified Hopkinson Bars (/s).

Laminate	Carbon/ Glass Type 1	Carbon/ Glass Type 2a	Carbon/ Glass Type 2b	All Carbon	All Glass
Original SHPB	1040	965	990		
Modified SHPB	1620	1600	1510	855	1120

positions on the failure plane. It is clear from these results that there will be a very marked effect of loading rate on the likelihood of delamination succeeding first tow failure and on the extent of delamination should it occur. As stated in Section 2.4, attempts are still being made to develop a test configuration for which the variation in shear strain on the failure plane is less marked. The present test arrangement, however, is considered adequate for investigating the general effect of strain rate on the interlaminar shear properties and, more especially, for obtaining data on the behaviour of the particular laminates used in the previous tensile impact tests. This will require the fabrication of specimens to the design given in Figure 6 but using the same wet hand lay-up procedure, the same epoxy resin and reinforced with the same woven carbon/epoxy and woven glass/epoxy fabrics used for the tensile specimens and including a hybrid lay-up such that the interlaminar failure occurs at the interface between a carbon-reinforced and a glass-reinforced ply. This work is currently in progress.

6. CONCLUSIONS

Progress has been made in both the experimental and the theoretical areas of the research programme. On the experimental side, in addition to the continuing programme of tensile impact tests on hybrid carbon/glass and carbon/Kevlar laminates, progress has also been made in developing an experimental technique for determining the interlaminar shear properties of laminated composites at different rates of strain. Initial results for a woven carbon/epoxy laminate at nominal shear strain rates of 0.0003/s and 500/s show a very marked increase in the interlaminar shear strength and an equally marked reduction in the nominal interlaminar shear strain at failure under impact loading.

Theoretical work has progressed in three specific areas, i.e., studies of the stress distribution around the specimen/holding-bar interface, further work on the prediction of hybrid tensile impact strength using laminate theory and the development of a finite element method for studying the tensile failure process in woven hybrid laminates. The main conclusions are outlined below.

(1) In an attempt to minimise stress concentrations at the interface between the specimen and the loading-bar, and hence to optimise the load transfer to the specimen in both quasi-static and impact tests, the ABAQUS finite element package was used to study the effect of changes in the loading-bar geometry on the stresses close to the interface. In practice, none of the four modified loading-bar geometries studied showed any

significant reduction in the stress concentrations at the interface with the specimen so the standard design of bar with a rectangular end was retained. For an all-glass specimen, where the difference in elastic properties between the loading-bar and the specimen was greatest, significant concentration factors for both the shear and the normal stresses were only observed within about 1 mm of the interface.

(2) Laminate theory predictions of the hybrid quasi-static tensile strength have been extended to include the behaviour observed under impact loading and a reasonably good agreement with the experimentally determined strengths has been obtained. The failure strains are, however, significantly overestimated. Attempts to refine this analysis by assuming a range of tow failure strengths leading to successive single tow failures at increasing levels of applied load and taking into account the local stress concentrations around failed tows are, as yet, far from conclusive. The possibility has been demonstrated, however, of a hybrid effect arising from the ability of the glass-reinforced plies, following the failure of an individual carbon fibre tow, to reduce the stress concentrations in neighbouring carbon-reinforced plies.

(3) A finite element technique has been developed for analysing tensile failure processes in woven hybrid composites. This technique allows estimates to be made of the stress and strain distribution around a failed element. The normal and shear stress (or strain) concentration factors, determined in this way on the interlaminar plane, are likely to control the subsequent stages in the failure process. The effect of different stacking sequences and of initial failure in different reinforcing plies has been investigated and features observed in the experimental work highlighted. Further work is needed to characterise the interply shear and normal strengths before a full numerical analysis can be developed to model the actual failure of the specimen.

7. FUTURE WORK

The present report describes work performed during the first two years of a three-year programme entitled "Modelling of the Impact Response of Fibre-Reinforced Composites". It is anticipated that the final stages in both the experimental and the theoretical programmes will be as described below.

7.1 Experimental Programme

Apart from a few remaining tensile impact tests required to complete the test programme on carbon/glass and Kevlar/glass epoxy laminates, two main areas of experimental work are still to be completed. Both are concerned with the provision of data for use in the finite element studies of the fracture process in woven hybrid laminates under impact tension. The first series of tests is to determine the effect of loading rate on the interlaminar shear strength for failure at the interface between (1) two carbon-reinforced plies, (2) two glass-reinforced plies and (3) a carbon-reinforced and a glass-reinforced ply. In initial tests on special specimens fabricated from woven carbon pre-preg, see Section 4.3, a very marked effect of strain rate was observed.

These tests now have to be performed on specimens fabricated from the specific woven carbon and woven glass fabrics and the same epoxy resin as was used in the fabrication of the original hybrid laminates. The method of fabrication of these specimens, wet hand lay-up, will try to reproduce as accurately as possible that used previously for the simple laminates. Tests will then be required at an impact and a quasi-static rate of loading.

In parallel with the interlaminar shear strength tests it is also required to determine the effects of strain rate on the tensile strength normal to the interlaminar plane for the same three types of interface. This will necessitate the fabrication of thicker laminates, with up to 24 carbon plies or as many as 50 glass plies. Using specimens cut from laminates of this thickness it should also be possible to determine the effect of strain rate on the laminate elastic properties in the direction normal to the plane of reinforcement. In all finite element analyses so far rough estimates have had to be made of the elastic constants in the thickness direction at both quasi-static and impact rates of strain.

7.2 Theoretical Studies

As described in detail in Chapter 2, finite element analyses have been performed for the strain distribution in woven hybrid laminate specimens following the failure of one or more plies for different positions of the point of first ply failure and for different hybrid lay-ups. The results obtained illustrate the general effects of hybridisation and of the specific hybrid lay-up on the type of failure that might be expected. For a more detailed description of the failure process, however, it is necessary to consider various possible processes whereby failure may propagate following the initial tensile failure of a given ply. In practice, the particular sequence of events following this initial failure is likely to be determined by the relation between the relative strength levels for (1) fibre failure in tension, (2) delamination under interlaminar shear and (3) deplying under tension normal to the interlaminar plane and the actual stress (or strain) field within the specimen due to the initial failure of a given ply. Once experimental data becomes available on the effect of strain rate on the interlaminar shear and interply normal failure strengths it will be possible to take further the analyses of Chapter 2 and, it is hoped, obtain a more detailed description of the processes controlling failure under impact tension.

While awaiting this experimental data it is also hoped to extend the analyses already reported on here to include some of the specific cases arising in the discussion of Section 5.1 and related to a refinement of the laminate theory approach. In particular it is planned to investigate the effect on the local stress and strain field of failure in one or more plies of an all carbon specimen and, possibly, using a three-dimensional analysis, of the failure also of one or more individual tows.

8. ACKNOWLEDGEMENT

This research was sponsored by the Air Force Office of Scientific Research, Air Force Systems Command, USAF, under Grant No. AFOSR-87-0129.

9. REFERENCES

1. Parry, T. and J. Harding. 1988. Colloque Int. du CNRS No. 319, *Plastic Behaviour of Anisotropic Solids*. J. P. Boehler, ed. Paris: CNRS, pp. 217–288.

2. Chiem, C. Y. and Z. G. Liu. 1988. *Proc. IMPACT '87, Impact Loading and Dynamic Behaviour of Materials*. C. Y. Chiem, H.-D. Kunze and L. W. Meyer, eds. Oberursel: DGM Informationsgesellschafft mbH, 2:579–586.

3. Campbell, J. D. and W. G. Ferguson. 1970. *Phil. Mag.*, 21:63–82.

4. Werner, S. M. and C. K. H. Dharan. 1986. *J. Comp. Mater.*, 20:365–374.

5. Harding, J., Y. L. Li, K. Saka and M. E. C. Taylor. 1988. *Proc. 4th. Oxford Int. Conf. on Mech. Props. of Materials at High Rates of Strain*. London and Bristol: Institute of Physics.

6. Saka, K. and J. Harding. 1986. "Behaviour of Fibre-Reinforced Composites under Dynamic Tension", Interim Report on Grant No. AFOSR-85-0218 (Oxford University Engineering Laboratory Report, No. OUEL 1654/86).

7. Saka, K. and J. Harding. 1985. "Behaviour of Fibre-Reinforced Composites under Dynamic Tension", Final Report on Grant No. AFOSR-84-0092 (Oxford University Engineering Laboratory Report, No. OUEL 1602/85).

8. Shah, S., R. K. Y. Li and J. Harding. 1988. "Modelling of the Impact Response of Fibre-Reinforced Composites", Interim Report on Grant No. AFOSR-87-0129 (Oxford University Engineering Laboratory Report, No. OUEL 1730/88).

9. Tsai, S. W. and E. M. Wu. 1971. *J. Comp. Mater.*, 5:58–80.

CHAPTER 2

Failure Analysis of Woven Hybrid Composite Using a Finite Element Method

Y. Li, C. Ruiz and J. Harding

ABSTRACT: A finite element method is used to determine the stress and strain distributions around a failure link in a carbon-reinforced ply in a woven carbon/glass laminate under tensile loading. On the assumption that delamination follows this initial tensile failure in a carbon link, the new stress and strain distribution is determined. Finally, the effect of stacking sequence on the failure of a woven hybrid laminate is discussed.

1. INTRODUCTION

Hybrid composites have many advantages, for example, they offer an effective way of increasing the impact strength and reducing the cost of an advanced composite material. In addition, the crack arresting properties and the fracture toughness of the composite can be improved due to the 'hybrid effect' in fibre fracture strain.

In the general case, a hybrid composite consists of two kinds of fibres. One fibre has a lower critical fracture strain, i.e., the low elongation (LE) phase, the other fibre has a high critical fracture strain, i.e., the high elongation (HE) phase. In this model, the LE phase is the carbon ply and the HE phase is the glass ply. If the bond between the fibre and the resin is perfect, the strain in the model is compatible, i.e., the HE phase and the LE phase have the same tensile strain provided that a uniform strain is applied to the specimen. The LE phase is first broken when the uniform strain is larger than the critical fracture strain of the LE phase. After, the LE phase breaks a tensile strain concentration and a shear stress concentration is produced around the position of the breakage. The tensile strain concentration may lead to the failure of neighbouring plies. On the other hand, the shear stress

55

concentration may result in a delamination between the failure ply and neighbouring plies of the HE phase. Which type of failure occurs following the break of the LE ply depends on the ratio of maximum tensile strain to the ultimate tensile strain and the ratio of the maximum shear stress to the ultimate interlaminar shear stress.

In this report, the initial strain and stress redistribution and strain concentration factor around the failed carbon ply are calculated, using ABAQUS, for an interply woven hybrid laminate of carbon and glass introducing a discontinuity in the modulus to model failure. Assuming delamination follows the break of this ply, the strain redistribution is also determined. Finally, the effect of stacking sequence of carbon and glass plies on the fracture process is discussed.

2. EXPERIMENTAL AND NUMERICAL TECHNIQUES

2.1 The Finite Element Method

The finite element method has found wide application in composite material research since Puppo and Evensen [1] first used it for studying the interlaminar shear in laminated composites in 1970. Typical applications include the 'boundary effect problem' in laminated composites [2–4] and modelling the fracture process in unidirectional fibre composites [5]. Generally speaking, both the micro-mechanical approach and the macro-mechanical approach can be used for calculating the stress or strain distribution.

In the micro-mechanical approach, the fibres and resin are treated independently and the geometry of the fibre and the interaction between fibre and resin, and fibre and fibre can be considered. This means that the model has to be divided into a large number of elements and a computer with high speed and large storage is needed. In the macro-mechanical approach, the micro-structure of the material is not considered, and the composite material is treated as orthotropic [6]. In view of the particularly complicated structure of the woven reinforcement geometry we use the macro-mechanical approach as a first attempt to obtain a numerical analysis solution of this problem.

2.2 Experimental Detail

Tensile specimens, as shown in Figure 1, have been tested in a split Hopkinson-bar device as described in [7]. The quasi-static elastic constants

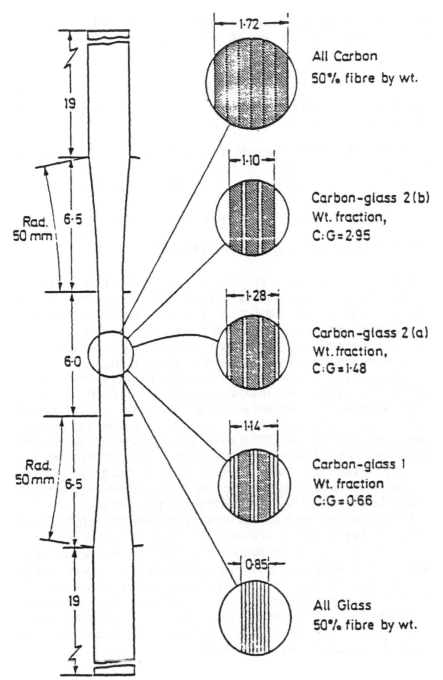

FIGURE 1. General arrangement of tensile specimen.

for non-hybrid specimen reinforced with the same glass or the same carbon plies were measured by K. Saka [8]. The mean values were:

	E_1 (GPa)	E_2 (GPa)	ν_{13}	ν_{31}	ν_{31}^*
All-glass	16.6	13.8	0.17	0.14	0.14
All-carbon	45.3	43.3	0.14	0.09	0.13

The Poisson's ratio ν_{31}^* determined from symmetry hypothesis of an orthotropic laminate, i.e., $E_1\nu_{31} = E_3\nu_{31}$, is also shown. The subscripts 1, 3 correspond to the longitudinal direction and to the transversal direction respectively (See Figure 2).

A detailed description of the experimental work is contained in [8–10].

2.3 Finite Element Analysis of Hybrid Specimen

The stress distribution in the standard design of tensile specimen has been determined using the PAFEC finite element package for both an all-glass and a hybrid carbon/glass lay-up. In each case the tensile stress in the specimen gauge region was significantly higher than the tensile or shear stresses elsewhere in the specimen leading to the conclusion that the specimen design is satisfactory and that an initial tensile failure, as normally obtained in practice, is what would be expected. The use of the ABAQUS finite element package with triangular elements made possible a more accurate modelling of the waisted geometry of the specimen and, for an all-glass lay-up, confirmed the above conclusion regarding the specimen design. Since then the ABAQUS analysis has been extended to allow a study of the stress distribution in a hybrid specimen with the particular aim of estimating the magnitude of any stress concentrations arising at points on the interface between the carbon and the glass-reinforced plies where they intersect the free surface in the waisted region of the specimen.

The finite element mesh used for the analysis of the type 2b hybrid specimen, including the loading bar to which it is attached, is shown in Figure 2. The hybrid lay-up consists of alternating carbon- and glass-reinforced plies with a total of two glass and three carbon plies in the cross section of the central parallel region, where the stress is assumed to be uniform. The main purpose of the stress analysis was to check the validity of this assumption rather than to obtain accurate value of the stresses. To this end, approximate value of the elastic constants were taken, as shown in Figure 2. The material was treated as orthotropic, with

$$E_1 = E_2 \quad \nu_{13} = \nu_{31}$$

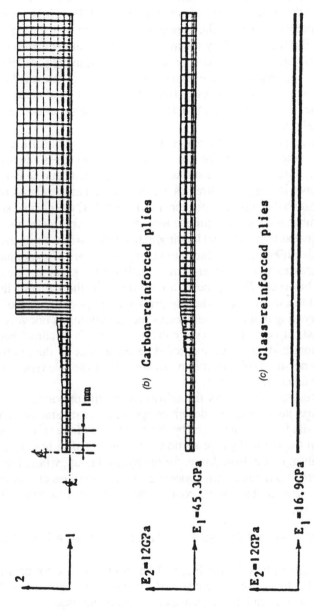

(a) Specimen and loading-bar

1 mm

(b) Carbon-reinforced plies

$E_2=12GPa$

$E_1=45.3GPa$

(c) Glass-reinforced plies

$E_2=12GPa$

$E_1=16.9GPa$

FIGURE 2. Finite element mesh for hybrid carbon/glass tensile specimen.

The value taken for the tensile modulus in direction 2 is based on the rule of mixture, assuming a fibre volume fraction of 50%. ν_{12} was taken to be equal to 0.15. In the absence of any experimental data, it is not possible to provide precise value to the elastic properties in the through-thickness direction but this is unlikely to have much significant effect on the conclusion. The longitudinal tensile stresses (direction 1) in the parallel gauge section of the specimen were found to exceed the normal stresses (direction 2) and the interfacial shear stresses (on the 1-3 plane) by some five orders of magnitude, confirming that this part of the specimen is effectively in a state of uniaxial tension.

The corresponding stress states in the tapered region of the specimen, along the interface between the carbon- and the glass-reinforced plies where they intersect the free surface, positions A-B-C and D-E-F in Figure 3, are shown in Figure 4. Stresses are determined at the Gauss integration points close to the nodes. Along the interface A-B-C the normal stresses are three or more orders of magnitude less and the shear stresses two orders of magnitude less than the local longitudinal stresses while along the interface D-E-F this difference is reduced to about two orders of magnitude for the normal stresses and one order of magnitude for the shear stresses. Close to point D, however, discrepancies are apparent in the value of the normal stress and the shear stress when determined on opposite sides of the interface, suggesting that for these stresses the calculation is not very accurate. In contrast, at both interfaces a good agreement is obtained between the value of longitudinal stress measured on either side of the interface when the difference in elastic moduli in direction 1 for the two types of reinforcing ply is taken into account.

These results confirm, for the hybrid lay-up, the previous conclusion for the all-glass lay-up that the design of specimen is satisfactory and that an initial tensile failure determined by the magnitude of the ruling longitudinal stresses in the parallel gauge section, as is observed in practice, is the expected failure mode. In addition, for the hybrid lay-up, where there are distinct differences in the elastic properties of the two types of reinforcing ply, it is now shown that stress concentration at the free surface remains relatively small.

2.4 Finite Element Analysis of a Specimen with a Failed Link

The specimen type 2(a) in Figure 1 has been selected for detailed analysis. In order to ascertain the effect of the stacking sequence on the failure process, several stacking sequences have been analysed:

1. G-C-G-C-G-C-G (Basic alternating sequence)
2. C-G-G-C-G-G-C
3. G-G-C-C-C-G-G

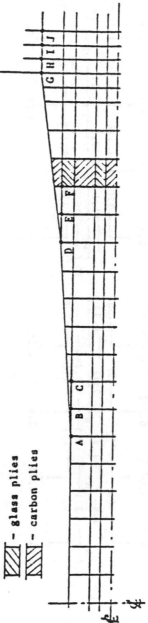

- glass plies
- carbon plies

FIGURE 3. Waisted region of hybrid tensile specimen.

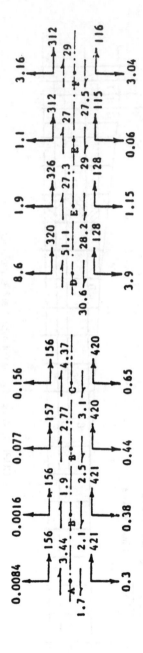

Interface A-B-C　　　　　　**Interface D-E-F**

FIGURE 4.　Stress near free surface in waisted region of hybrid tensile specimen.

Since the stresses away from the central reduced thickness region of the specimen are relatively small and the stress concentration over the tapered transition are negligible, it is possible to simplify the numerical analysis by considering the central region. Figure 5 shows the general dimensions of mathematical model. A state of plane strain is assumed and a mesh consisting of 210 elements is used as shown in Figure 6. Each element has a length of 0.2 mm. Assuming that the first stage in the fracture process is the tensile failure of one of the carbon fibre tows, as is observed in the practice, then to determine whether the next stage is also controlled by the longitudinal stresses or whether a delamination mechanism related to the local shear stresses or a deplying mechanism under the local normal stresses becomes the controlling process, it is necessary to determine the redistributed stress system in the vicinity of first failure. The carbon and the glass-reinforced plies are divided into a number of elements, or links, and first failure is assumed to occur in an arbitrarily chosen link in one of the carbon-reinforced plies.

To model the tensile failure of the link the longitudinal modulus is reduced by a factor of 0.001. This leads to a singularity at each node of the failed link and high shear stresses on the interface with the neighbouring plies at points close to these nodes. However, discontinuities in the normal and shear stresses across the interface close to the singularity indicate that here again the analysis has become inaccurate and the stress levels determined cannot be relied upon. This makes difficult the identification of the next stage in the failure process from a comparison of those stresses with the critical stresses for a tensile, a normal or a shear failure at the given strain rate.

Figure 7(a) shows a section of the specimen, comparing elements in two adjacent plies, one reinforced with carbon, the other with glass. When a uniform strain is applied to the specimen, the gauge lines AB and CD that define the two sections, move to A'B' and C'D'. If one of the plies breaks in the specimen, the specimen will respond by deforming in a non-uniform manner, so that the gauge lines will distort as shown in Figure 7(b). The interface of two adjacent plies are still subjected to the same tensile strain but the shear strains are different, being in the ratio of the shear moduli. The finite element analysis provides values of the stress and strain at points such as P and Q by extrapolation from the Gauss points. To characterize the state of strain in each ply and the shear stress at the interface, we note that:

- The tensile strain is the same either of the interface between adjacent plies.
- The shear stress is continuous across the interface between adjacent plies.

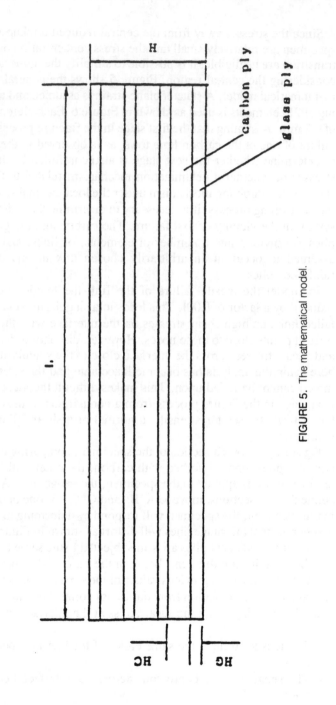

FIGURE 5. The mathematical model.

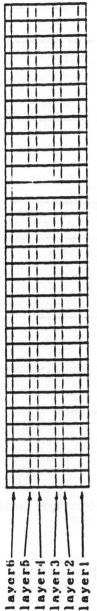

FIGURE 6. The mesh of finite element and the definition of layers.

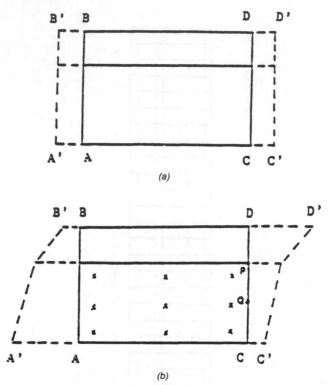

FIGURE 7. The state of strain in neighbouring carbon- and glass-reinforced plies.

The tensile strain and the shear stress between adjacent plies are calculated in the report.

3. RESULTS

3.1 Failure of a Central Carbon Ply in the Basic Alternating Stacking Sequence (Figures 8, 9 and 10)

In the experimental work, it was found [9] that at a quasi-static rate, the mean failure strain for the all-carbon lay-up was 1.35% ± 0.3% and for all-glass lay-up 2.52% ± 0.2%. In the analysis, therefore, it was assumed that the first carbon ply failed when an overall strain of 1.35% was applied.

In this stress analysis a perfect bond between neighbouring plies, i.e., compatibility of strain between the two types of ply, has been assumed. The corresponding distorted mesh for the failure of a link in the central carbon ply when an overall strain of 1.35% is applied to the parallel gauge region of specimen is shown in Figure 8(c). If the resulting tensile strain concentration in the neighbouring glass- or carbon-reinforced plies [Figure 8(a)], is sufficiently high their tensile failure may be the next stage in the failure process. Alternatively the shear concentration on the interlaminar plane between the failed ply and the neighbouring reinforced plies [Figure 8(b)], could lead to delamination as the next stage in the failure process. Which type of failure follows is determined, therefore, by the relative values of the critical tensile strain to the critical shear stress. Estimates of the former are available from the tensile tests on the non-hybrid carbon- and glass-reinforced laminates. The methods of determining the critical condition for interlaminar shear strength are still being investigated.

Experimental evidence suggests that, for woven reinforced laminates, a limited delamination follows tensile failure of a given fibre tow, the extent of delamination being greater under impact loading. In the model being considered, if delamination is represented simply as a crack between two frictionless surfaces the shear stress concentration at the singularity is not reduced very much by the delamination process but merely moves with the delamination crack tip, i.e., the delamination extends catastrophically through to the ends of the specimen. While such behaviour is sometimes observed, particularly in tests on unidirectionally-reinforced specimens, it does not describe the response seen in the present woven hybrid composites. Here, therefore, a friction coefficient of 0.5 is assumed on the delamination crack surfaces between which slip is allowed to occur. The delamination crack is modelled using the ABAQUS interface element for which, since there should be no relative movement until slip occurs, a modulus of 1000 GPa is assumed.

The resulting distorted mesh shapes, for delamination extending over either two or five links on either side of the initially failed link, and the relative tensile strain and shear stress distributions are shown in Figures 9 and 10. On the basis of these calculations delamination does lead to a reduction in the peak shear stress on the interlaminar surface at or near the delamination crack tip, by about 25.5% for a 2-link delamination [Figure 9(b)] and by about 31.1% for a 5-link delamination [Figure 10(b)]. This is not a large reduction. Whether it is significant or not may depend on the balance of probability between tensile and shear failure in any given case and on how sensitive shear failure is to the very localised shear concentration. As shown in Figure 8(b), even without delamination the shear stress at the interface has fallen by more than an order of magnitude within 5 links, i.e., within 1 mm, of the singularity.

TENSILE STRAIN DISTRIBUTION

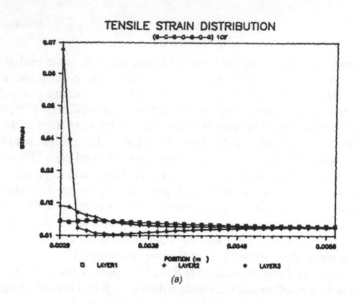

(a)

SHEAR STRESS DISTRIBUTION

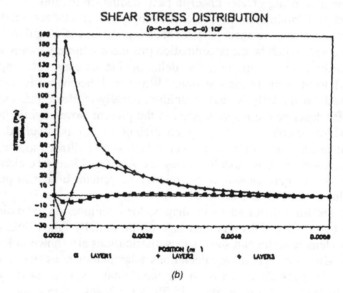

(b)

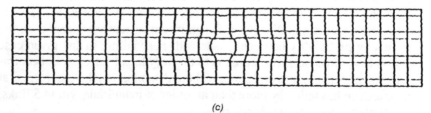

(c)

FIGURE 8.

TENSILE STRAIN DISTRIBUTION
(0-C-0-0-0-0-0) 1CF+0

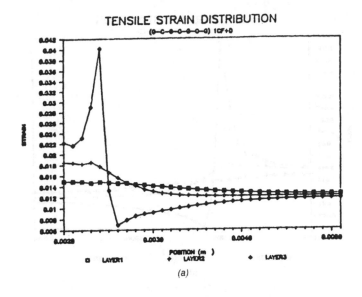

(a)

SHEAR STRESS DISTRIBUTION
(0-C-0-C-0-C-0) 1CF+0

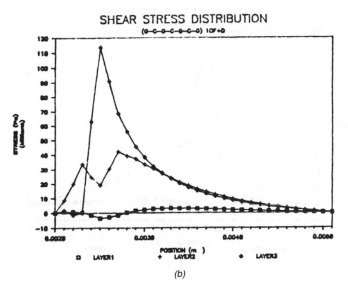

(b)

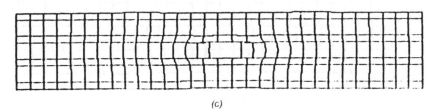

(c)

FIGURE 9.

69

TENSILE STRAIN DISTRIBUTION

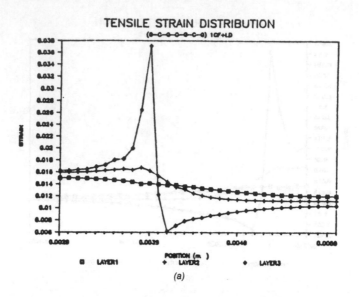

(a)

SHEAR STRESS DISTRIBUTION

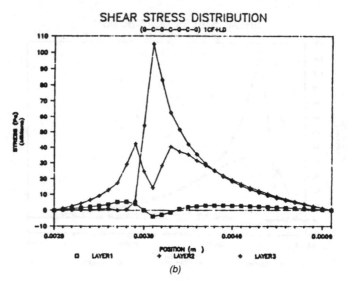

(b)

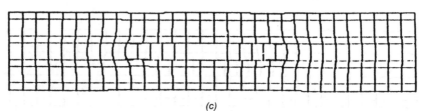

(c)

FIGURE 10.

3.2 Failure of a Lateral Carbon Link in the Basic Alternating Sequence (Figures 11, 12, and 13)

The same technique has been followed to model this type of failure. Due to the lack of symmetry, bending occurs in the specimen and the peak tensile strain and shear stress are slightly higher than before. Delamination is particularly noticeable in Figure 13.

3.3 Failure of a Glass Link in the Basic Alternating Sequence (Figures 14 and 15)

Although this is a very implausible situation, it could occur if, for example, one of the glass-reinforced plies is defective. Figure 14 shows the results of a failure in one of the innermost glass plies. It is obvious that the tensile strain is very much smaller than that found when the carbon ply fails and that the shear stress is also very small. The same conclusion is reached when it is the outermost glass ply that fails as shown in Figure 15.

3.4 Failure of all Carbon Plies in the Basic Alternating Sequence (Figures 16 and 17)

If failure of central carbon ply at an average strain of 13.5% is followed by failure of the neighbouring outer carbon plies at the same strain without any delamination, a peak tensile strain is reached in the remaining glass plies of about 10%, see Figure 16. These plies may then break in tension. On the other hand, very high shear stresses are also present, which may cause delamination between the failed links and the glass plies. If this happens, the tensile strain in the glass plies is then relieved, dropping from 10% to 5.5%. The interlaminar shear stresses are also reduced, from 250 MPa to 130 MPa, as shown in Figure 17. It follows that the final failure could be either fracture across the whole specimen with or without limited pull-out (delamination) or by tensile failure of the carbon plies and pull out of the glass plies, depending on the relative tensile strength and failure strain of the glass plies and the interlaminar shear strength of carbon/glass interface.

3.5 Other Stacking Sequences (Figures 18–25)

For comparison with the basic stacking sequence, the following stacking sequences and failures have also been analysed. The results are shown in the figures indicated. Each lay-up has the same overall proportion of carbon to glass (3 carbon plies to 4 glass plies).

Sequence	Failure	Figure No.
G-G-C-C-C-G-G	Central C	18
G-G-C-C-C-G-G	Central C + delamination	19
G-G-C-C-C-G-G	All C	20
G-G-C-C-C-G-G	All C + delamination	21
C-G-G-C-G-G-C	Central C	22
C-G-G-C-G-G-C	Central C + delamination	23
C-G-G-C-G-G-C	All C	24
C-G-G-C-G-G-C	All C + delamination	25

4. SUMMARY OF RESULTS AND DISCUSSION

4.1 Failure Process

It follows from these results that the strain and stress fields will vary throughout the volume of the tensile specimen following the initial (primary) failure. The shear stress is produced by the fibre failure. Both tensile strain and shear stress concentrations occur in the ply adjacent to the breakage. Which kind of failure (ply fracture or delamination) follows the initial failure depends on the critical value of shear stress and tensile strain. It is important, therefore, to obtain an estimate of the interlaminar shear strength experimentally before the present study can be taken much further.

In practice, experimental results show that the critical tensile fracture strain of the LE (carbon) phase is governed by a statistical distribution and so all the fibres of the LE phase will not fail at the same level of strain although with the present woven reinforcement configuration, the statistical variation in LE ply failure strain is likely to be much less significant. A region of pseudo-yield behaviour is observed in which the specimens are slowly damaged provided the volume of HE fibre (glass) is enough to support the increased loading. This means that the fibres do not fail in the specimen one after another in catastrophic succession. The tensile strain concentration may be reduced during this process as a result of delamination, since, if delamination occurs, the strain concentration will be relaxed. This, therefore, will be considered in the next section. Otherwise, if the failure is still by ply fracture, final catastrophic failure will follow very quickly, and the pseudo-yield behaviour will not be observed in the experiment.

If shear stress is the limiting parameter, delamination will occur after the first ply failure. As a result of the delamination, fast fracture of all the plies will be prevented because delamination will reduce the tensile strain concentration in the ply adjacent to the first break. Table 1 shows the maximum

TENSILE STRAIN DISTRIBUTION
(O—C—C—C—C—O) TCF

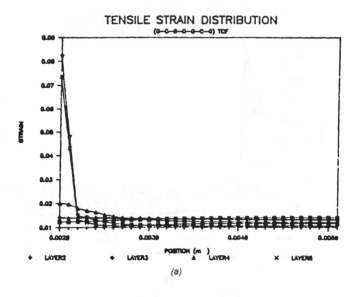

(a)

SHEAR STRESS DISTRIBUTION
(O—C—C—C—C—O) TCF

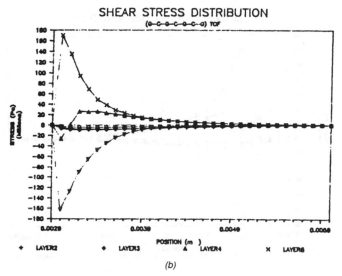

(b)

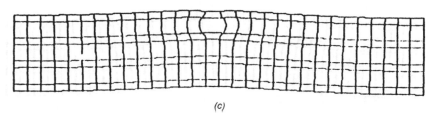

(c)

FIGURE 11.

TENSILE STRAIN DISTRIBUTION

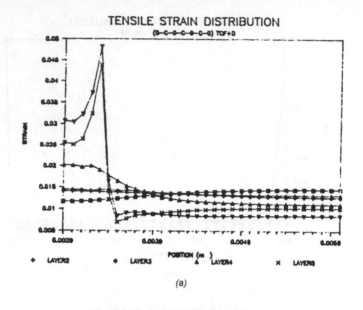

(a)

SHEAR STRESS DISTRIBUTION

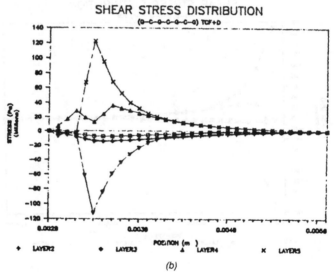

(b)

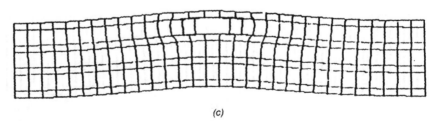

(c)

FIGURE 12.

74

TENSILE STRAIN DISTRIBUTION

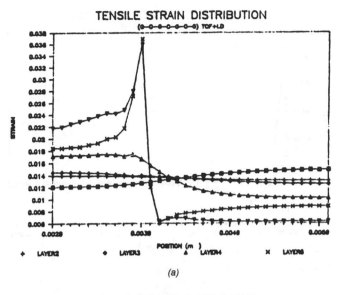

(a)

SHEAR STRESS DISTRIBUTION

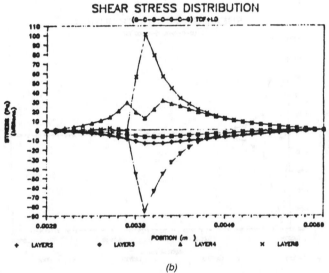

(b)

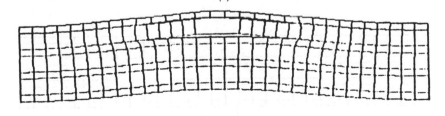

(c)

FIGURE 13.

75

TENSILE STRAIN DISTRIBUTION

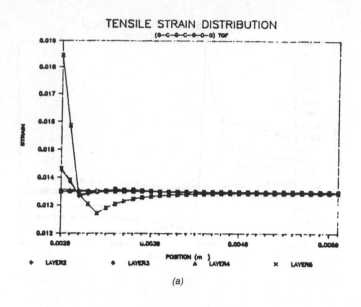

(a)

SHEAR STRESS DISTRIBUTION

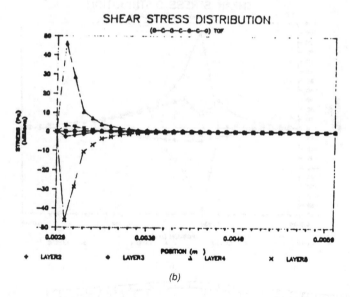

(b)

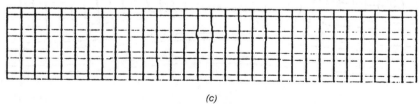

(c)

FIGURE 14.

TENSILE STRAIN DISTRIBUTION

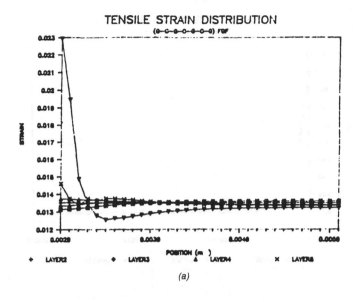

(a)

SHEAR STRESS DISTRIBUTION

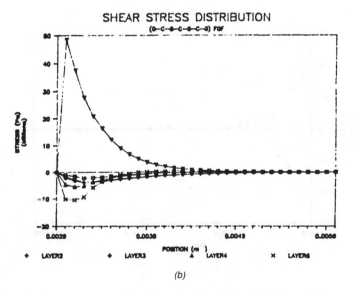

(b)

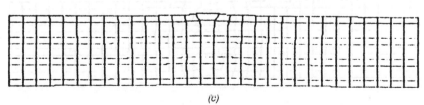

(c)

FIGURE 15.

77

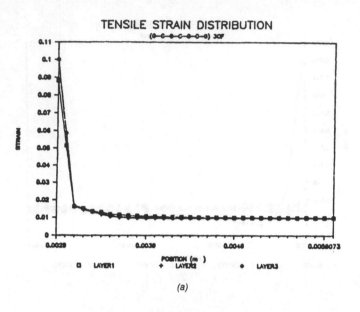

TENSILE STRAIN DISTRIBUTION

(a)

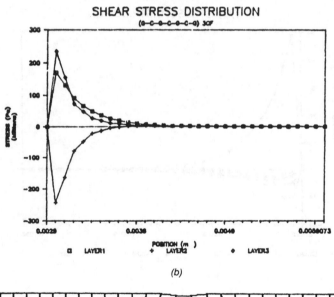

SHEAR STRESS DISTRIBUTION

(b)

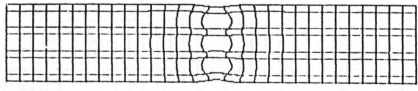

(c)

FIGURE 16.

TENSILE STRAIN DISTRIBUTION

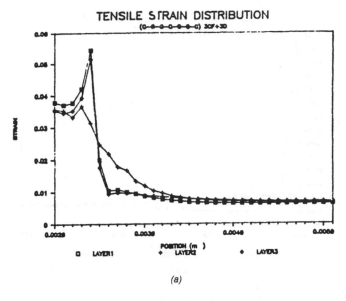

(a)

SHEAR STRESS DISTRIBUTION

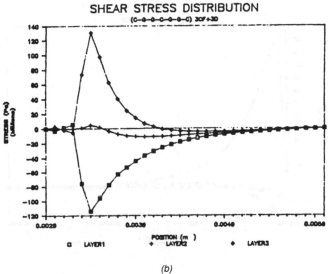

(b)

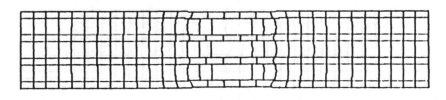

(c)

FIGURE 17.

TENSILE STRAIN DISTRIBUTION

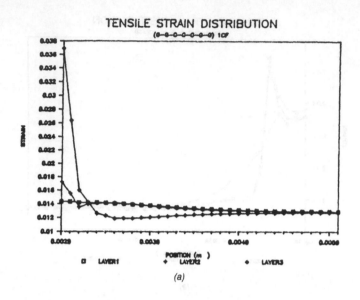

(a)

SHEAR STRESS DISTRIBUTION

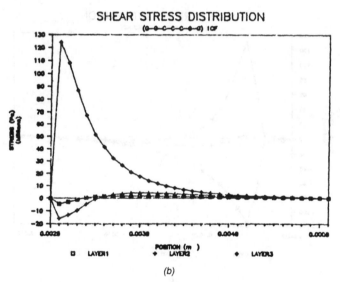

(b)

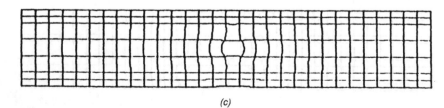

(c)

FIGURE 18.

80

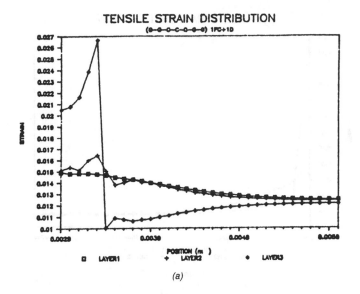

(a)

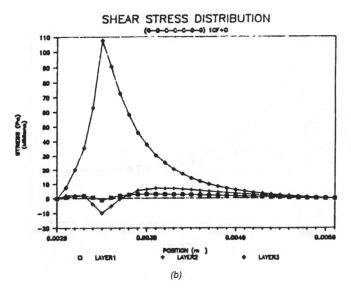

(b)

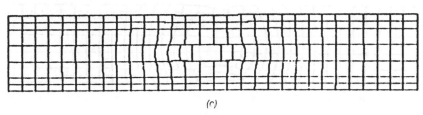

(c)

FIGURE 19.

81

TENSILE STRAIN DISTRIBUTION

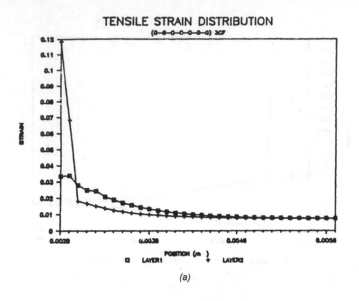

(a)

SHEAR STRESS DISTRIBUTION

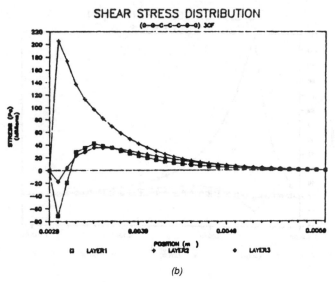

(b)

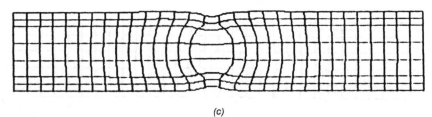

(c)

FIGURE 20.

TENSILE STRAIN DISTRIBUTION
(0—0—0—0—0—0—0) 3CF+3D

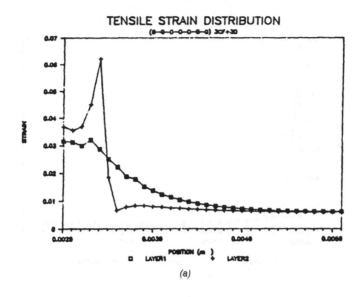

(a)

SHEAR STRESS DISTRIBUTION
(0—0—0—0—0—0—0) 3CF+3D

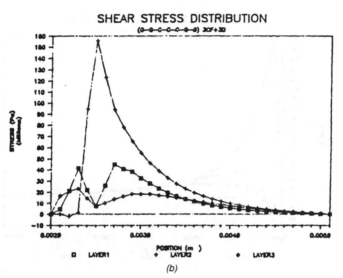

(b)

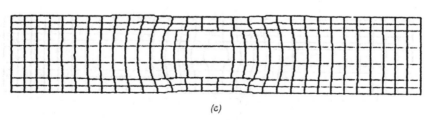

(c)

FIGURE 21.

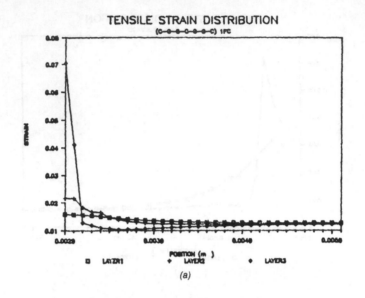

TENSILE STRAIN DISTRIBUTION

(a)

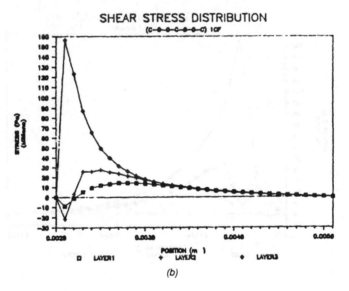

SHEAR STRESS DISTRIBUTION

(b)

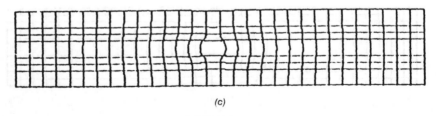

(c)

FIGURE 22.

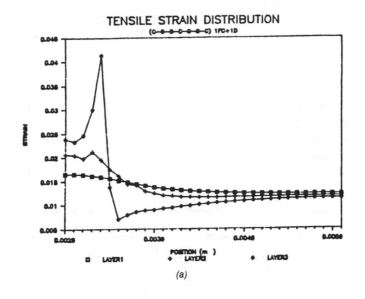

(a)

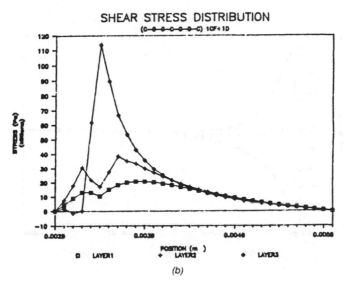

(b)

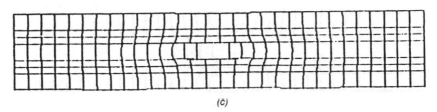

(c)

FIGURE 23.

TENSILE STRAIN DISTRIBUTION

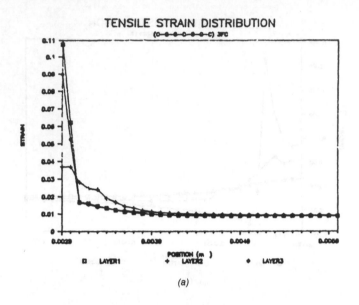

(a)

SHEAR STRESS DISTRIBUTION

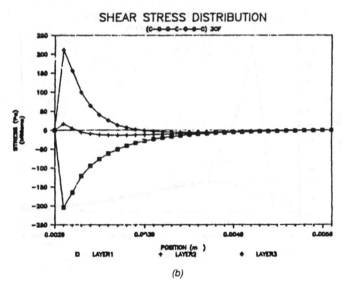

(b)

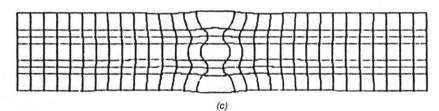

(c)

FIGURE 24.

86

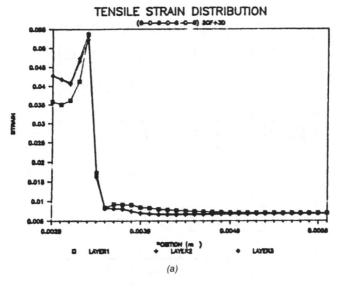

(a)

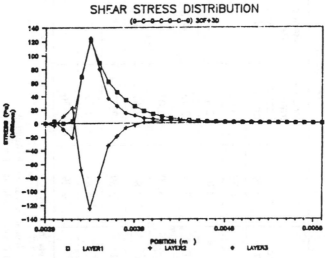

(b)

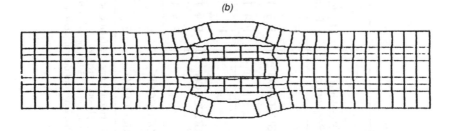

(c)

FIGURE 25.

TABLE 1.

Stacking Sequence	Failure Form	ε_{11max} (%)	Concentration Factor on ε_{11}	Reduced Percentage of ε_{11}	τ_{12max} (MPa)	Reduced Percentage of τ_{12}	Fig. No.
GCGCGCG	F	6.78	5.02		152.6		8
	F + D	4.01	2.96	40.4	113.63	25.5	9
	F + LD	3.71	2.75	45.1	105.1	31.1	10
GCGCGCG	F	8.20	6.07		170.3		11
	F + D	4.81	3.57	41.2	122.46	28.1	12
	F + LD	3.69	2.73	55.0	101.01	40.7	13
GCGCGCG	F	1.84	1.37		46.3		14
GCGCGCG	F	2.29	1.70		48.4		15

C, G: indicates that the failure occurs in this ply.
F: one ply only breaks.
D: the length of delamination is 0.4 mm at each side of broken link.
LD: the length of delamination is 1.0 mm at each side of broken link.

value of tensile strain and shear stress in the adjacent ply after delamination. Note that in Figure 9, the delamination length is 0.4 mm, i.e., 2 elements at each side of the broken link, and in Figure 10 it is 1 mm, i.e., 5 elements.

Comparing Figure 8 to Figures 9 and 10, it is found that delamination reduces the tensile strain concentration by 41% while the shear stress concentration is reduced by about 25.5%. This implies that tensile fibre fracture may be avoided once delamination takes place. If delamination propagates further, the maximum tensile strain and shear stress change only a little. This shows that the tensile strain concentration will be relaxed mainly at the early stage of delamination. Similarly Figures 12 and 13 can be compared to Figure 11. From Figures 8 to 13, it can also be seen that, when the initial failure occurs in a carbon ply, the tensile strain concentration in the nearest carbon ply after delamination, about 2.74%, is the same whichever carbon ply fails. So the nearest carbon ply is also likely to break if the loading increases continuously.

Following the initial carbon ply failure, delamination takes place. Then, the nearest carbon ply fails, repeating this process until all the carbon plies break provided that the glass plies are still able to stand the increased loading. Finally, there is fast fracture of all glass plies once the tensile strain exceeds their critical fracture strain. Delamination in the failed specimens has been observed most markedly in tests on unidirectionally-reinforced composites and, to a lesser extent, in experimental results on woven hybrid composites of carbon and glass to which this report refers. Although the tensile strain concentration in Figure 11 is larger than that in Figure 9, the maximum value of the tensile strain after delamination, about 3.7%, is nearly the same, as shown in Figures 10 and 13. Again, the tensile strain concentration is reduced quickly during the early stages as delamination propagates. The tensile strain concentration finally disappears if delamination propagates through the whole length of the specimen. The reduction of tensile strain in the ply adjacent to the failed ply is related to the length of delamination and also to the maximum value before delamination.

4.2 Stacking Sequence

In terms of the laminate theory, the stacking sequence does not affect the initial elastic constants which only depend on the total fibre volume and on the volume fraction of the two types of fibre. The elastic constants can be predicted by the rule of mixtures (ROM). This is verified by the experimental results of many investigators [10]. For strength, an extension of the laminate theory approach in an attempt to predict the strength properties of fabric-reinforced hybrids has been made by Saka [8]. He used the Tsai-Wu criterion and laminate theory to predict the strength of hybrid carbon/glass laminates with different volume fractions. The results of his analytical predictions were in slightly better agreement with experimental results than those given by the ROM. Unlike the elastic constants, the strength of the hybrid composite material depends on many factors of which stacking sequence is one of the most important. It is difficult, therefore, to predict the strength of the hybrid composite only using the volume fraction of the constituents.

Following the principles of fracture mechanics, it may be accepted that crack growth occurs when the amount of strain energy released per unit area equals the energy absorbed in the creation of free surfaces. If the potential energy is U, the principle is expressed in the form,

$$\frac{dU}{da} = G \tag{1}$$

where G is taken to be a property of the material. In the specimen that has been studied, the potential energy corresponding to an overall strain of 1.35% over a gauge length of 6 mm is 24.7 J for the unit width, regardless of the stacking sequence. The energy released when the first carbon ply breaks, depends on the stacking sequence, as shown below:

Type	Stacking sequence	$\Delta U(J)$ for first carbon ply failure
1	G-C-G-C-G-C-G	1.6 (no delamination)
		2.7 (delamination)
2	G-G-C-C-C-G-G	1.2 (no delamination)
		2.2 (delamination)
3	C-G-G-C-G-G-C	1.2 (no delamination)
		2.8 (delamination)

The higher the amount of energy released, the easier it will be to satisfy the condition for crack growth [Equation (1)]. From those results it follows that the stacking sequence type 1 is more likely to exhibit a break in the central carbon ply than the other two although it is less likely to delaminate than type 3.

Extending this analysis to the case of a failure in all the carbon plies, the results obtained are:

Type	Stacking sequence	$\Delta U(J)$ for failure of all carbon plies
1	G-C-G-C-G-C-G	6.6 (no delamination)
		12.0 (delamination)

Type	Stacking sequence	$\Delta U(J)$ for failure of all carbon plies
2	G-G-C-C-C-G-G	11.4 (no delamination)
		12.9 (delamination)
3	C-G-G-C-G-G-C	8.3 (no delamination)
		2.8 (delamination)

Ignoring the possibility of delamination, it is now seen that type 1 can only release 6.6 J of potential energy while type 2 releases 11.4 J, making it the most likely to fail through crack growth in the carbon plies. Type 3 is intermediate between the two, at 8.3 J. The same conclusion is reached if the delamination is considered. It is worth noting that the number of delaminations is 6 in type 1, 2 in type 2 and 4 in type 3. The energy released per delamination is comparatively small.

The conclusion is that the alternating stacking sequence (type 1) will tend to start breaking before the other two, but because the total amount of potential energy that it can release is so low when more carbon plies fail, it will need more external work to produce the final failure.

The preceding treatment does not pretend to model the exact process of failure, but it still serves to highlight the difference between the three stacking sequences. Further development is obviously needed.

It is also interesting to apply the same approach to the alternating sequence in order to assess the most likely position of the first failure. The energy released for the failure of the central carbon ply, as we have seen, is 1.6 J (no delamination) and 2.7 J (delamination). If the outer carbon ply breaks, the corresponding values are 1.8 J and 4.1 J. It follows that an outer ply failure is the most likely with a relatively large amount of energy being released for the delamination process.

5. CONCLUSIONS

It is quite clear that the failure of the specimen may be initiated by the tensile fracture of a carbon ply. Whether this is in the centre or towards the surface, depends on statistical considerations since all plies are subjected to the same nominal strain. If the central ply breaks, the tensile strain in the adjacent glass ply increases by a factor of 5.02 and a maximum shear stress equal to 152.6 MPa appears between the broken ply and the adjacent glass plies. The maximum tensile strain in the carbon-reinforced plies also increases but by a small amount. What happens next depends on the shear strength of the interply surface and on the tensile strength of the neighbouring glass ply. If delamination follows the first break, the tensile strain falls and so, to a lower extent, does the shear stress. This implies that delamination does not necessarily extend over the whole specimen. If the glass ply breaks following the fracture of the carbon, a fast tensile fracture results. Experimental observations tend to support the view that delamination plays an important role in determining the mode of failure.

When an outer carbon ply breaks, the situation is very similar except that the maximum tensile strain in the glass increases now by a factor of 6.07 and the maximum shear stress becomes 170.3 MPa. This means that, if a discontinuity is unavoidable in the manufacturing process, it is better to put it in the centre. Delamination has the same effect as before.

The hybrid effect not only depends on the volume fraction of two types of fibres, but also on the stacking sequence. If the volume fraction of two fibres is constant, the stacking sequence should affect the amount of energy released for carbon plies failure of laminate. The initial stiffness is not related to the stacking sequence.

The finite element method has been shown to highlight the features found in the experimental work. Further work is needed to characterize the interply shear strength before a full numerical analysis can be developed to model the actual failure of specimen. The importance of the stacking sequence on the hybrid effect is also clear from the variation in strain concentration and shear strain values depending on the order in which tensile fracture and delamination occur.

6. REFERENCES

1. Puppo, A. H. and H. A. Evensen. 1970. "Interlaminar Shear in Laminated Composite under Generalized Plane Stress", *J. of Composite Materials*, 4:204–220.

2. Reddy, J. N. and D. Sandidge. 1978. "Mixed Finite Element Models for Laminated Composite Plates", *Trans. ASME, J. of Engineering for Industry*, 109:39–45.

3. Pipes, R. B. and N. J. Pagano. 1970. "Interlaminar Stress in Composite Laminated under Uniform Axial Extension", *J. of Composite Materials*, 4:538–548.

4. Wong, C. M. S. and F. L. Matthews. 1981. "A Finite Element Analysis of Single and Two-Hole Bolted Joints in Fiber Reinforced Plastic", *J. of Composite Materials*, 15:481.

5. Mandel, J. M., S. C. Pack and S. Tarazi. 1982. "Micromechanical Study of Crack Growth in Fiber Reinforced Material", *Engineering Fracture Mechanics*, 16(5):741–754.

6. Shah, S., R. K. Y. Li and J. Harding. 1987. "Modelling of the Impact Response of Fiber-Reinforced Composite", *O.U.E.L. Report*, No. 1706/87.

7. Harding, J. and L. M. Welsh. 1983. "A Tensile Testing Technique for Fiber Reinforced Composites at Impact Rates of Strain", *J. of Mater. Sci.*, 18:1810–1826.

8. Saka, K. 1987. "Dynamic Mechanical Properties of Fiber-Reinforced Plastics", D. Phil. Thesis, Department of Engineering Science, Oxford University.

9. Harding, J. and K. Saka. 1986. "Behaviour of Fibre-Reinforced Composites under Dynamic Tension—Third Progress Report", *O.U.E.L. Report*, No. 1654/86.

10. Harding, J. and K. Saka. 1988. "The Effect of Strain Rate on the Tensile Failure of Woven-Reinforced Carbon/Glass Hybrid Composites", in *Proc. IMPACT '87* (DGM Informationsgesellschafft mbh, Oberursel), 1:515–522.

CHAPTER 3

Characterisation of the Impact Strength of Woven Carbon Fibre/Epoxy Laminates

J. Harding, Y. L. Li, K. Saka and M. E. C. Taylor

ABSTRACT: Techniques for determining the impact mechanical response of laminated composites in tension, compression and interlaminar shear are briefly described. Stress-strain curves for a plain weave carbon epoxy laminate under both tensile and compressive impact are compared with those obtained at a quasi-static rate and interlaminar shear strengths are determined at both rates for a satin weave carbon/epoxy laminate.

1. INTRODUCTION

In view of their highly anisotropic nature, the development of a failure criterion for laminated composites usually requires the determination of the characteristic strengths of the particular reinforcing plies in more than one direction and under more than one type of loading. Thus the commonly used Tsai-Wu criterion may require a knowledge of the tensile, compressive and in-plane shear strengths in two orthogonal directions. In addition to these in-plane failure modes, however, laminated composites may also fail either by delamination under a critical interlaminar shear stress or by deplying under a critical stress normal to the interlaminar plane. The relative importance of these various possible failure processes will depend on the type and rate of loading and on the geometry, reinforcement configuration and ply lay-up in the given composite laminate. As a first step towards a fuller understanding of which parameters most affect the composite impact response, therefore, the present paper describes techniques for determining the tensile, compressive and interlaminar shear strengths under impact loading for two woven carbon/glass epoxy laminates and compares the results obtained with those from quasi-static tests.

95

2. SPECIMEN MATERIALS

Tension and compression specimens were prepared from a plain weave carbon-epoxy laminate supplied by Fothergill and Harvey. The reinforcing fabric was woven from Toray 3000 filament fibre tows, type T300–3000A, and had a weight of 189g/m and an approximate thickness of 0.28 mm. The fabric had a relatively coarse weave geometry with only 47 ends and picks per 10 cm. The laminate was hand laid-up using Ciba-Geigy XD927 epoxy resin, with 100 parts by weight of resin to 36 parts by weight of hardener and a cure schedule of 24 hours at room temperature followed by 16 hours at 100°C. The final laminate contained 12 layers of fabric and had a fibre content of 50% by weight and a thickness of 3.4 mm. For both types of loading a thin strip specimen, waisted in the thickness direction, as shown in Figure 1(a), was used. A finite element analysis [6] of the stress distribution in this specimen showed an essentially uniaxial stress system in the central parallel region, the stresses being significantly higher here than elsewhere in the specimen. In practice it was found that both tensile and compressive failures almost always occurred in this region.

Several designs of interlaminar shear specimen were investigated, that finally chosen being shown in Figure 1(b). Specimens to this design could not be cut from the same pre-cured woven carbon/epoxy laminate as the tension and compression test specimens and so had to be fabricated from a woven carbon/epoxy pre-preg available in the laboratory. This employed a 5-end satin weave fabric, woven from 3000 filament fibre tows with, respectively, 70 and 72 yarns per 10 cm in the warp and weft directions and having a dry weight of 285 g/m^2. The pre-preg was manufactured by Hexcel and Genin using a type ES.36 self-adhesive epoxy resin to give a fibre weight fraction of 52% and an uncured pre-preg weight of 548 g/m^2. Eight layers of pre-preg were laid up, using metal and PTFE spacers as shown in Figure 1(b), and then covered by a sealing sheet and evacuated to 28 inches of mercury. This was then placed inside a closed container at an air pressure of 90 psi and the container placed inside a small oven. The temperature was raised over a period of an hour to 125° C, held for 2 hours and then allowed to cool to room temperature.

In view of the different starting materials and fabrication routes a direct comparison between the results of the interlaminar shear tests, on the one hand, and the tension and compression tests, on the other, is not possible. A finite element analysis of the stress distribution in the interlaminar shear specimen (see Figure 2) shows large shear strain (and hence shear stress) concentrations at each end of the failure plane, points "X" in Figure 2. This is a major drawback of this design of specimen and means that the measured load at failure gives only an average value for the shear stress on the

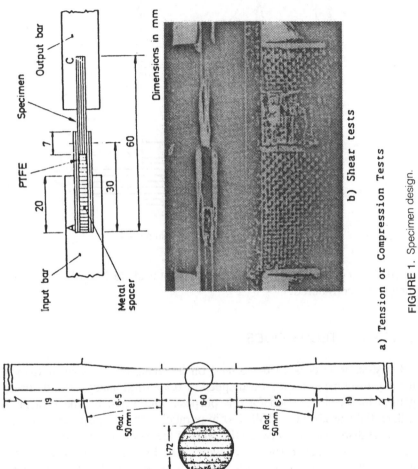

Dimensions in mm

a) Tension or Compression Tests

b) Shear tests

FIGURE 1. Specimen design.

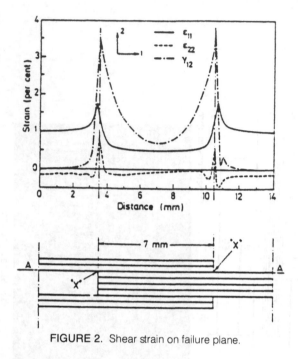

FIGURE 2. Shear strain on failure plane.

interlaminar plane. This problem was apparent in all designs of shear specimen so far considered.

3. TESTING TECHNIQUES

Quasi-static tests were performed on a standard screw-driven Instron loading machine at a crosshead speed of 0.2 in/min using a small strain-gauged load cell designed for this purpose, and a pair of linear variable differential transformers (LVDT's) in parallel with the specimen to determine the displacement between the ends of the loading bars. For the compression tests a special rig was designed (see Figure 3) to ensure that the loading bars were accurately aligned and to prevent buckling of the specimen/loading bar/load cell system between the fixed and moving crossheads. For the tension and compression tests strain gauges attached to either side of the specimen parallel gauge region were used to monitor the elastic strain in the specimen. The load cell, LVDT voltage-time and, where applicable, specimen strain gauge signals, were stored in Datalabs' type DL902 transient recorders and subsequently analysed in an IBM PC. Typical test records for a tension test on a specimen loaded in the weft direction are shown in Figure 4 and the corresponding stress-strain curves are given in Figure 5. The specimen strain gauges are used to correct the initial part of the stress-strain curve as derived from the displacement transducers.

All impact tests were performed using the Hopkinson-bar technique where the specimen is fixed between two long elastic loading bars. For both the tension and the shear tests a tensile stress wave was propagated along the input bar using a special gas-gun and loading system described elsewhere [4]. For the compression tests a small air-gun and a direct impact loading system was used. Voltage-time signals from the two strain gauge stations on the input bar, the one strain gauge station on the output bar and, for the tension and compression tests, the strain gauges on the specimen, were stored in two Datalabs type 912 high-speed transient recorders and were subsequently analysed using an IBM PC. Typical test records for a compression test on a specimen loaded, in this case, in the warp direction are shown in

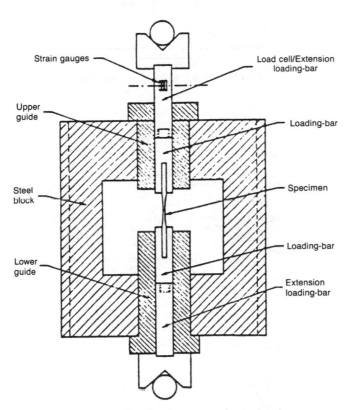

FIGURE 3. Quasi-static compression testing rig.

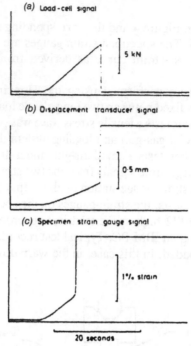

FIGURE 4. Raw data for quasi-static test.

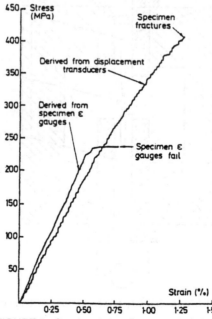

FIGURE 5. Quasi-static tensile stress-strain curves.

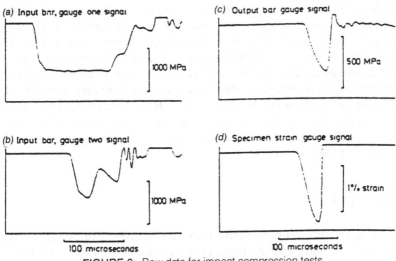

FIGURE 6. Raw data for impact compression tests.

Figure 6. The corresponding stress-strain curves for strains derived from (1) the Hopkinson-bar analysis, which does not give an accurate measure of the strain during the early stages of the test, and (2) the specimen strain gauges, which only give a measure of the elastic part of the test, are compared in Figure 7. The Hopkinson-bar analysis is seen to overestimate the strain in the initial region by about 0.5%. The specimen strain gauges are used to calibrate the strain measurement in the elastic region, giving the final corrected stress-strain curve, shown by the dashed line in Figure 7.

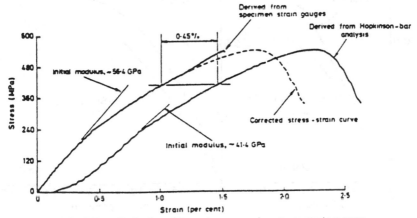

FIGURE 7. Derivation of impact compression stress-strain curves.

In the tests on the shear specimens the applied load is transmitted as a shear force across the interface between neighbouring reinforcing plies, interfaces A-A in Figure 2. The corresponding shear zone width is both very narrow and not clearly defined. Using the standard Hopkinson-bar analysis, displacements are determined, during the course of the test, between the adjacent ends of the two loading bars. However, it has not yet been possible to determine how much of this overall displacement actually appears across the shear zones at interfaces A-A so the corresponding shear strains cannot be calculated and only the load supported by the specimen during the course of the test is determined.

4. RESULTS

Mean stress-strain curves, derived from a minimum of three such tests at both the quasi-static and the impact rates of strain, are compared for the tension tests in Figure 8(a) and the compression tests in Figure 8(b). For these tests all specimens were loaded in the weft direction. Mean values of the elastic modulus, the maximum stress and the failure strain and estimates of the "yield" stress and strain, corresponding to the limit of the linear elastic response, are given in Table 1. The experimental scatter band is indicated in each case. A comparison of the failure stresses in the quasi-static and the impact shear tests is given in Table 2.

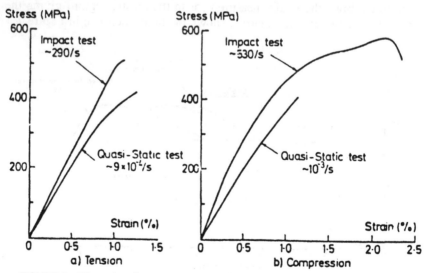

FIGURE 8. Effect of strain rate on stress-strain curves in tension and compression.

TABLE 1. Effect of Strain Rate on the Laminate Mechanical Properties.

Strain Rate	Modulus (GPa)	Maximum Stress (MPa)	Failure Strain (%)	"Yield" Stress (MPa)	"Yield" Strain (%)
		Tensile Loading			
Quasi-static	42.5 ± 3%	416 ± 4%	1.24 ± 8%	172 ± 13%	0.40 ± 14%
Impact	49.0 ± 8%	508 ± 2%	1.10 ± 9%	482 ± 5%	1.00 ± 14%
		Compressive Loading			
Quasi-static	39.9 ± 7%	400 ± 7%	1.06 ± 3%	185 ± 9%	0.47 ± 14%
Impact	62.5 ± 4%	584 ± 3%	2.20 ± 19%	156 ± 14%	0.24 ± 8%

The fracture appearance of specimens tested in tension, compression and shear is shown in Figures 9(a), 9(b) and 1(b), respectively, for tests at impact rates of loading. Little effect of loading rate was observed on the failure mode. In both the tension and the compression tests failure is seen to initiate in the central parallel region of the specimen, where the stress system was essentially uniaxial and parallel to the direction of loading, while in the shear test failure was on the two expected interlaminar planes. In calculating the interlaminar shear strengths given in Table 2 it is assumed that failure occurs at the same load on both these planes. In view of the relatively coarse weave of the carbon fibre reinforcement, and hence the far from planar interlaminar failure surface, this assumption may be only approximately true.

5. DISCUSSION

The testing technique used here for determining the in-plane tensile impact properties of composite laminates is now well established and reliable results have been obtained for materials reinforced both with different fibre types [7] and with hybrid reinforcements [5]. In impact compression, however, all previous studies have used short cylindrical specimens. This is not an ideal geometry for testing anisotropic two phase materials. The present investigation is the first to use the waisted strip specimen, as recommended for quasi-static compression testing of fibre-reinforced composites [2]. While the experimental scatter observed in these tests (see Table 1) was generally quite small, more work is required before complete confidence can be placed in the interpretation of the test data. In particular, the mean stress-strain curve in Figure 8(b) masks some differences in the post "yield" response shown by the three specimens tested, differences which have yet to be related to differences in fracture behaviour. The shear test is still in the early stages of development and only average values for the shear stress on the interlaminar plane at failure are reported. Nevertheless, despite the non-uniform shear stress distribution on this plane, the present technique has the advantages of a less complex specimen geometry and a better defined stress system than in previous test configurations [1,3]. Even so,

TABLE 2. *Effect of Strain Rate on the Interlaminar Shear Strength.*

Strain Rate	Shear Strength
Quasi-static	33.4 ± 2.0 MPa
Impact	54.3 ± 0.9 MPa

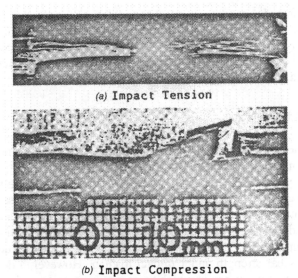

(a) Impact Tension

(b) Impact Compression

FIGURE 9. Failure modes under impact loading.

modifications to the present design of shear specimen and loading system aimed at giving a more uniform stress distribution on the failure plane are still being considered.

As is apparent from Figure 8 and from Tables 1 and 2 the experimental results show a marked effect of strain rate on the mechanical response for all three loading configurations. However, while both the elastic modulus and the failure strength at the quasi-static rate have very similar values in both tension and compression, under impact loading they show a significantly greater increase in compression than in tension. Also, while there is a marginal decrease in the tensile failure strain under impact loading, the compressive failure strain is more than doubled. In consequence the stress-strain response under impact compression cannot strictly be described as linear-elastic to failure. This may inhibit the direct use of the compressive failure strength so determined in failure criteria such as that of Tsai-Wu. The increase in interlaminar shear strength with strain rate, about 60% (see Table 2) is even more marked than that shown in either tension or compression, where the increases were about 20% and 45%, respectively. The tensile failure strength is likely to be controlled primarily by fibre fracture whereas an examination in the optical microscope of the failure surface in the shear tests gave no indication of fibre fracture. In the compressive failure process, although fibre fracture is clearly observed [see Figure 9(b)], the matrix may well play a more important role than for tensile failure. It so, the difference in the rate dependence of the failure strengths in tension, compression and shear may be related, qualitatively, to the differing importance of the rate dependent matrix properties in resisting failure. In view, however, of the different laminates used in the tension and compression tests, on the one hand, and the shear tests, on the other, a direct comparison of the rate dependence in each case is perhaps not valid.

6. CONCLUSIONS

Techniques have been developed for the impact testing of composite laminates in tension, compression and interlaminar shear. Results for woven carbon fibre/epoxy laminates show significant increases in the tensile and compressive failure strengths with strain rate and some increase in the tensile and compressive elastic moduli. While the failure strain is marginally reduced under impact loading in tension it is almost doubled in compression. A significant increase is also observed in the average value of the interlaminar shear strength under impact loading. However, a design of interlaminar shear specimen in which there is a more uniform distribution of shear stress on the failure plane is still being sought.

7. ACKNOWLEDGEMENT

This research was sponsored by the Air Force Office of Scientific Research, Air Force Systems Command, USAF, under Grant No. AFOSR-87-0129.

8. REFERENCES

1. Chiem, C. Y. and Z. G. Liu. 1988. *Proc. IMPACT '87* (DGM Informationsgesellschafft mbH, Oberursel), 2:579–586.

2. Ewins, P. D. 1971. R.A.E. Technical Report No. 71217.

3. Parry, T. and J. Harding. 1988. Colloque Int. du CNRS No. 319, *Plastic Behaviour of Anisotropic Solids*. Paris: CNRS, pp. 271–288.

4. Saka, K. and J. Harding. 1985a. Oxford University Engineering Laboratory, Report No. OUEL 1602/85.

5. Saka, K. and J. Harding. 1985b. *Proc. IUTAM Colloquium on Macro- and Micro-Mechanics of High Velocity Deformation and Fracture, Tokyo*. Berlin: Springer-Verlag, pp. 97–111.

6. Shah, S., R. K. Y. Li and J. Harding. 1988. Oxford University Engineering Laboratory, Report No. OUEL 1730/88.

7. Welsh, L. M. and J. Harding. 1985. *Proc. DYMAT '85*. Paris: Jour de Physique, Colloque C5, 405–414.

CHAPTER 4

A Simple Laminate Theory Approach to the Prediction of the Tensile Impact Strength of Woven Hybrid Composites

K. Saka and J. Harding

1. INTRODUCTION

With the recent development [1] of a version of the Hopkinson-bar apparatus suitable for the tensile impact testing of composite materials several successful attempts have been made [2,3] to obtain fundamental information on the effect of strain rate on the deformation and fracture mechanisms in both unidirectionally and woven-reinforced polymer composites. These studies have shown a much greater increase in strength and impact resistance for glass/epoxy composites under impact loading than for carbon/epoxy composites. More recently this work has been extended to include hybrid lay-ups of woven glass and woven carbon in an attempt to determine those parameters which optimise the composite mechanical response under impact loading [4].

Considerable data have been generated as a result of these various investigations so that reliable experimental information is now available on the tensile impact response of a range of fibre-reinforced composites. It becomes possible for the first time, therefore, to attempt an extension of the techniques so far used only to model the quasi-static behaviour of such materials to allow a description of their behaviour under impact loading. This paper is concerned with a simple laminate theory approach to the modelling of the tensile impact response of woven-reinforced hybrid carbon-glass/epoxy laminates, in terms of the experimental data obtained on woven carbon/epoxy and woven glass/epoxy laminates in the various studies described above. These data have been obtained using the split Hopkinson-bar apparatus in which it is assumed that the specimen, although not the loading system, will be in equilibrium during all but the initial stages of the test. Effects of stress wave propagation within the specimen, therefore, may be ignored in the present analysis, although there is evidence that such effects may be of considerable significance in the impact of larger laminated structures [5].

2. APPLICATION OF LAMINATE THEORY TO HYBRID COMPOSITES

2.1 Stress Analysis

Classical laminate theory is based on the generalised form of Hooke's Law,

$$\sigma_{ij} = C_{ijkl}\,\epsilon_{kl} \qquad (i,j,k,l = 1,2,3) \tag{1}$$

where σ_{ij} is the second rank stress tensor, ϵ_{kl} is the second rank strain tensor and C_{ijkl} is the fourth rank stiffness tensor. For a two-dimensional orthotropic plate, where 1 and 2 are the principal material directions, this simplifies to

$$\begin{Bmatrix} \sigma_1 \\ \sigma_2 \\ \tau_{12} \end{Bmatrix} = \begin{bmatrix} Q_{11} & Q_{12} & 0 \\ Q_{12} & Q_{22} & 0 \\ 0 & 0 & Q_{66} \end{bmatrix} \begin{Bmatrix} \epsilon_1 \\ \epsilon_2 \\ \gamma_{12} \end{Bmatrix} \tag{2}$$

where

$$Q_{11} = E_1/(1 - \nu_{12}\nu_{21}); \quad Q_{22} = E_2/(1 - \nu_{12}\nu_{21}); \quad Q_{66} = G_{12} = G_{21}$$

and

$$Q_{12} = \nu_{12}E_2/(1 - \nu_{12}\nu_{21}) = \nu_{21}E_1/(1 - \nu_{12}\nu_{21}) \tag{3}$$

The elastic constants E_1, E_2, ν_{12}, ν_{21} and G_{12} may be determined experimentally for each type of reinforcing ply in the given hybrid lay-up at the appropriate rate of loading [6].

The constitutive equation for laminated two-dimensional orthotropic composite materials may be written as

$$\begin{Bmatrix} N \\ M \end{Bmatrix} = \begin{bmatrix} A & B \\ B & D \end{bmatrix} \begin{Bmatrix} \epsilon^0 \\ \chi \end{Bmatrix} \tag{4}$$

where N is the resultant tractions, M is the resultant moments, ϵ^0 the midplane strains, χ is the plate curvatures and A, B and D are the extensional stiffness matrix, the coupling stiffness matrix and the bending stiffness matrix, respectively. The matrix A is defined as

$$A_{ij} = \sum_{k=1}^{n} (\bar{Q}_{ij})_k (h_k - h_{k-1}) \tag{5}$$

where (Q_k) is the $[\bar{Q}]$ matrix of the kth layer of the laminate and h_k and h_{k-1} are defined in Figure 1 for a laminate composed of n layers. For a symmetrically stacked laminate the coupling stiffness matrix B is zero so from Equation (4) we have

$$\{\epsilon_j^0\} = [A_{ij}^{-1}]\{N_i\} \tag{6}$$

and, since $(Q) = (\bar{Q})$, there being no relative orientation between the principal material directions of adjacent plies,

$$A_{ij} = \sum_{k=1}^{n} (Q_{ij})_k (h_k - h_{k-1}) \tag{7}$$

Considering an interlaminated symmetric composite with two different plain weave fabrics (designated a and b) as reinforcement where, of the n plies in the laminate, m are of type b fabric and h_a and h_b are the thicknesses of the constituent plies, then the extensional stiffness matrix, A_{ij}, is given by

$$A_{ij} = (n - m)h_a(Q)_a + mh_b(Q)_b \tag{8}$$

which may be written as

$$A_{ij} = (n - m)h_a \begin{bmatrix} (1 + \alpha\beta)(Q_{11})_a & (1 + \alpha\lambda)(Q_{12})_a & 0 \\ (1 + \alpha\lambda)(Q_{12})_a & (1 + \alpha\mu)(Q_{22})_a & 0 \\ 0 & 0 & (1 + \alpha\eta)(Q_{66})_a \end{bmatrix} \tag{9}$$

where parameters α, β, λ, μ and η are given by

$$\alpha = [m/(n - m)](h_b/h_a), \quad \beta - (Q_{11})_b/(Q_{11})_a, \quad \lambda = (Q_{12})_b/(Q_{12})_a \tag{10}$$

$$\mu = (Q_{22})_b/(Q_{22})_a \text{ and } \eta = (Q_{66})_b/(Q_{66})_a$$

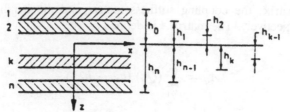

FIGURE 1. Definition of stacking parameters.

For uniaxial tension with the load applied in the x-direction, Equation (6) may be written,

$$
\begin{Bmatrix}
\epsilon_x^0 \\
\epsilon_y^0 \\
\gamma_{xy}^0
\end{Bmatrix}
= A_{ij}^{-1}
\begin{Bmatrix}
N_x \\
0 \\
0
\end{Bmatrix}
\tag{11}
$$

$$
= 1/[\phi(n - m)h_a]
\begin{Bmatrix}
(1 + \alpha\mu)(Q_{22})_a N_x \\
-(1 + \alpha\lambda)(Q_{12})_a N_x \\
0
\end{Bmatrix}
\tag{12}
$$

giving for the stress distribution in the carbon-reinforcing plies

$$
\begin{Bmatrix}
\sigma_x \\
\sigma_y \\
\tau_{xy}
\end{Bmatrix}_a
= 1/[\phi(n - m)h_a]
\begin{Bmatrix}
[(1 + \alpha\mu)(Q_{11})_a(Q_{22})_a - (1 + \alpha\lambda)(Q_{12})_a^2]N_x \\
[\alpha(\mu - \lambda)(Q_{12})_a(Q_{22})_a]N_x \\
0
\end{Bmatrix}
\tag{13}
$$

and in the glass-reinforcing plies

$$
\begin{Bmatrix}
\sigma_x \\
\sigma_y \\
\tau_{xy}
\end{Bmatrix}_b
= 1/[\phi(n - m)h_a]
\begin{Bmatrix}
[(1 + \alpha\mu)(Q_{22})_a(Q_{11})_b - (1 + \alpha\lambda)(Q_{12})_a(Q_{12})_b]N_x \\
[(1 + \alpha\mu)(Q_{22})_a(Q_{12})_b - (1 + \alpha\lambda)(Q_{12})_a(Q_{22})_b]N_x \\
0
\end{Bmatrix}
$$

$$
\tag{14}
$$

Thus, for a given applied load, i.e., traction N_x, Equations (13) and (14) may be used to determine the two-dimensional stress distribution in each type of reinforcing ply.

2.2 Failure Strength Criterion

Based on the very general assumption that a strength theory has to be of the form

$$f(\sigma_1, \sigma_2, \sigma_3) = 0 \qquad (15)$$

where σ_1, σ_2, and σ_3 are the principal stresses, Gol'denblat and Kopnov [7] proposed the criterion

$$(F_i \sigma_1)^\alpha + (F_{ij} \sigma_i \sigma_j)^\beta + (F_{ijk} \sigma_i \sigma_j \sigma_k)^\gamma + \ldots = 1 \qquad (16)$$

where F_i, F_{ij} and F_{ijk}, etc., are strength tensors, α, β and γ, etc., are material constants and $i, j, k = 1, 2, \ldots, 6$. Although Equation (16) may contain as many terms as may be desired it is usual to take only the first two and to assume specific values for α and β such that the criterion becomes

$$F_i \sigma_i + (F_{ij} \sigma_i \sigma_j)^{0.5} = 1 \qquad (17)$$

This has been found to give satisfactory results for a variety of orthotropic materials [7,8].

An operationally simple strength criterion, essentially similar to that of Gol'denblat and Kopnov, has been proposed by Tsai and Wu [9]. This has the form

$$F_i \sigma_i + F_{ij} \sigma_i \sigma_j = 1 \qquad (18)$$

where F_i and F_{ij} are defined as above. Since this criterion obeys tensor transformation rules it is especially suitable for engineering applications. The equivalence of Equations (17) and (18) in practice is supported by the work of Owen and Rice [10] who obtained indistinguishable results when the two criteria were applied to the biaxial strength of fabric-reinforced composites under both static and fatigue loading. It is also significant that in a review of multiaxial strength criteria for composite materials, Sendeckj [11] concluded that the Gol'denblat-Kopnov and Tsai-Wu formulations gave the best fit to the experimental data.

In its two-dimensional form the Tsai-Wu criterion may be written

$$F_{11}\sigma_1^2 + F_{22}\sigma_2^2 + 2F_{12}\sigma_1\sigma_2 + F_{66}\sigma_6^2 + F_1\sigma_1 + F_2\sigma_2 + F_6\sigma_6 = 1 \quad (19)$$

where $F_{12} = F_{21}$, $\sigma_6 = \tau_{xy}$ and F_{11}, F_{22}, etc., are defined as

$$F_{11} = 1/XX'; \; F_{22} = 1/YY'; \; F_{66} = 1/SS'; \; F_1 = (1/X) - (1/X');$$

$$F_2 = (1/Y) - (1/Y') \text{ and } F_6 = (1/S) - (1/S') \quad (20)$$

and where X, X', Y and Y' are the tensile and compressive strengths in the longitudinal and transverse (warp and weft) directions, respectively, and S and S' are the in-plane shear strengths. The interaction stress coefficient F_{12} is difficult to obtain experimentally so the usual assumption, proposed by Tsai and Hahn [12], is made, i.e., that

$$F_{12} = -(1/2)(F_{11}F_{22})^{0.5} \quad (21)$$

This gives for the strength criterion the expression

$$(\sigma_1^2/XX') + (\sigma_2^2/YY') - (\sigma_1\sigma_2)/(XYX'Y')^{0.5} + (\sigma_6^2/SS')$$

$$+ \; \sigma_1(X' - X)/XX' + \sigma_2(Y' - Y)/YY'$$

$$+ \; \sigma_6(S' - S)/SS' = 1 \quad (22)$$

Since, from Equations (13) and (14), $\tau_{xy} \; (= \sigma_6)$ is zero for the geometry we are considering, the appropriate two-dimensional form of the Tsai-Wu failure criterion reduces to

$$(\sigma_1^2/XX') + (\sigma_2^2/YY') - (\sigma_1\sigma_2)/(XX'YY')^{0.5} + \sigma_1(X' - X)/XX'$$

$$+ \; \sigma_2(Y' - Y)/YY' = 1 \quad (23)$$

2.3 Prediction of Failure Strength

For a given tensile loading, i.e., traction N_x, the state of stress in each ply, $\sigma_1 = \sigma_x$ and $\sigma_2 = \sigma_y$, may be determined from Equations (13) and (14). Each ply is then subjected to a failure strength test, i.e., the appropriate values of stress are substituted into Equation (23), and as N_x is increased the first ply to fail is identified. If the corresponding critical traction is N_x^*, the overall laminate stress in the direction of the applied load is given by

$$\sigma_x^* = N_x^*/[mh_b + (n - m)h_a] \tag{24}$$

and the corresponding laminate strain, from Equation (12), by

$$\epsilon_x^* = [(1 + \alpha\mu)(Q_{22})_a N_x^*]/[\phi(n - m)h_a] \tag{25}$$

Equations (24) and (25) relate only to first-ply-failure (FPF). If this does not immediately lead to catastrophic failure of the composite as a whole then to continue the calculation it is necessary to assess how the damage in the "failed" plies affects their subsequent capacity to carry load. In practice, when applied to the quasi-static tests, the above analysis predicts FPF to be in the carbon plies in each of the three hybrid lay-ups under consideration (see next section). The post-FPF stress distribution in the carbon plies is likely to be very complex and may not be amenable to an analytical approach based on laminate theory. Following, therefore, the early work of Tsai and Azzi [13], later adapted to woven fabric composites by Ishikawa and Chou [14], the post-FPF mechanical response of the carbon plies has been modelled as reduced stiffness matrix, each element except Q_{22} being reduced by a factor of 100. This choice of 100 for the reduction factor is arbitrary but the final result is not greatly sensitive to the factor chosen as was seen when an alternative factor of 10 was used and the results compared.

The assumption of a reduced stiffness matrix for the carbon plies following FPF will lead to altered values for the parameters β, γ and ϕ and hence to a modified version of Equation (14) for the stress distribution in the glass plies. It is assumed that the carbon plies continue to support load but that the bulk of any increased load will be carried by the glass plies. Final failure follows when the increased stresses in the glass plies satisfy the appropriate form of the Tsai-Wu criterion, Equation (23). The final overall composite stress and strain may then be determined from the corresponding critical traction. This leads to a bilinear approximation to the actual hybrid stress-strain response and assumes a linear-elastic behaviour through to failure for both the carbon- and the glass-reinforcing plies in the non-hybrid lay-ups. This assumption is supported by the experimental measurements at quasi-static rates but is not true for the all-glass specimens at impact rates where a clearly marked knee effect is observed [6].

3. EXPERIMENTAL RESULTS

For the present paper the data used were obtained from tests at a quasi-static rate of about 0.001/s and an impact rate of about 1000/s. Thin strip specimens cut from plain-weave carbon or glass/epoxy laminates were loaded in either the warp or the weft direction. Full details of the experimental techniques used in both tension [15,16] and compression [17,18] and of the data obtained [4,6] have been given elsewhere. From the tensile tests the in-plane elastic properties, i.e., Young's moduli, E_1 and E_2, and Poisson's ratios, ν_{12} and ν_{21}, and the failure strengths, X and Y in Equation (23), interpreted as the maximum stress just prior to failure, have been determined. The corresponding failure strengths, X' and Y', were obtained from similar tests performed in compression. The mean elastic constants so determined are listed in Table 1. For comparison, ν_{21} is also determined from the symmetry hypothesis for an orthotropic laminate, i.e., $E_1\nu_{21} = E_2\nu_{12}$, in giving an alternative value, ν_{21}^*. The discrepancy between these two values for the carbon-reinforced plies is not unusual [19,20]. A non-conservative behaviour of some constituents of the composite is usually postulated as the cause, possibly in this case the micro-cracking of the epoxy resin during the 'elastic' loading of the woven CFRP specimens. It is common in such cases to use the ν_{21}^* value in determining the stiffness matrix.

Mean values of the measured tensile and compressive failure strengths at both rates of loading are listed in Table 2. It may be noted that under quasi-static loading the carbon-reinforced plies show similar strengths in tension and compression but that in tension the warp direction is the stronger while in compression it is the weft direction which is stronger. In contrast, for the glass-reinforced plies the compressive strength is between one-half and one-third of the tensile strength but the warp direction remains the stronger under both tension and compression. The validity of these very low compressive strengths for the glass-reinforced plies may be open to question. In practice, however, for the particular loading configuration of interest, the compressive stresses developed in the two types of reinforcing ply are so small that little difference is made to the predicted hybrid failure strengths if the compressive strengths in Table 2 are assumed to be the same as those determined in tension.

TABLE 1. Mean Elastic Constants for Carbon- and Glass-Reinforced Plies.

Strain Rate	Material	E_1 (GPa)	E_2 (GPa)	ν_{12}	ν_{21}	ν_{21}^*
Quasi-static	All-glass	16.6	13.8	0.17	0.14	0.14
	All-carbon	45.3	43.3	0.14	0.09	0.13
Impact	All-glass	24.0	17.1	0.18	0.16	0.13
	All-carbon	48.7	49.0	0.06	0.08	0.06

TABLE 2. Mean Tensile and Compressive Failure Strengths for
Carbon- and Glass-Reinforced Plies.

		Tensile Strength		Compressive Strength	
Strain Rate	Material	Warp X	Weft Y	Warp X'	Weft Y'
Quasi-static (~ 0.001/s)	All-glass All-carbon	328 MPa 428 MPa	262 MPa 416 MPa	129 MPa 373 MPa	114 MPa 400 MPa
Impact (~ 1000/s)	All-glass All-carbon	494 MPa 562 MPa	427 MPa 508 MPa	397 MPa 570 MPa	387 MPa 584 MPa

Under impact loading the failure strengths are increased significantly in all cases. The same general response is observed as under quasi-static loading except that for the glass-reinforced plies the compressive failure strengths, while still below, are now closer to the tensile strengths while for the carbon-reinforced plies the compressive failure strengths in both warp and weft directions exceed the corresponding tensile strengths.

Tensile tests have also been performed at both rates of loading on similar specimens cut from hybrid laminates reinforced with the same woven carbon and glass mats. Three different hybrid lay-ups, type 1 (GGCGGCGG), type 2a (GCGCGCG) and type 2b (CGCGC), were tested and the tensile failure strengths determined experimentally.

4. PREDICTION OF HYBRID TENSILE STRENGTH

4.1 Quasi-Static Loading Rate

When the appropriate values of the strength parameters, X, X', Y, and Y' in Table 2, are substituted into Equation (23) a quadratic is obtained in σ_1 and σ_2 with constant coefficients corresponding to failure in either the glass or the carbon plies. Using data from Table 1 stiffness matrices for the two types of reinforcing ply and hence values of the parameters β, γ and μ may be determined. Parameter α depends only on the particular hybrid lay-up. Hence, using Equations (13) and (14), the stress distribution in the two types of ply, in terms of the applied traction N_x may be determined. Table 3 gives the predicted stress distribution for loading in the warp direction for each of the three hybrid lay-ups.

TABLE 3. Stress Distribution in Carbon- and Glass-Reinforced Plies under Quasi-Static Loading in the Warp Direction.

Hybrid Lay-Up	Carbon-Reinforced Plies		Glass-Reinforced Plies	
	$\sigma_1/N_x(m^{-1})$	$\sigma_2/N_x(m^{-1})$	$\sigma_1/N_x(m^{-1})$	$\sigma_2/N_x(m^{-1})$
Type 1	1189	−8.2	437	6.6
Type 2a	962	−3.5	354	6.3
Type 2b	1053	−2.1	375	7.5

Thus, for the type 1 hybrid, FPF occurs in the carbon plies when the stresses

$$\sigma_1 = 1189 \, N_x \quad \text{and} \quad \sigma_2 = -8.2 \, N_x \qquad (26a,b)$$

first satisfy the appropriate version of Tsai-Wu, giving a critical traction of

$$N_x^* = 0.359 \, MN/m$$

or in the glass plies, when the corresponding stresses

$$\sigma_1 = 437 \, N_x \quad \text{and} \quad \sigma_2 = 6.6 \, N_x \qquad (27a,b)$$

give a critical traction of

$$N_x^* = 0.766 \, MN/m$$

Since the lower critical traction is obtained in the carbon-reinforced plies, these fail first at an overall composite stress and strain, given by Equations (24) and (25), of 276 MPa and 0.94 %. The actual stresses, σ_1 and σ_2, in the carbon and glass plies at FPF are found to be 427 and −8.2 MPa and 157 and 2.4 MPa respectively.

Following FPF a reduced stiffness matrix is assumed for the carbon-reinforced plies, giving modified values for the parameters β, λ and ϕ and a modified stress distribution in the glass-reinforced plies such that

$$\sigma_1 = 1354 \, N_x \quad \text{and} \quad \sigma_2 = 131 \, N_x \qquad (28a,b)$$

If the increase in traction between FPF in the carbon-reinforced plies and final failure of the composite, when the glass-reinforced plies also fail, is given by ΔN_x, then the total stress system in the glass-reinforced plies at failure is

$$\sigma_1 = 157 + 1354 \; \Delta N_x \quad \text{and} \quad \sigma_2 = 2.4 + 131 \; \Delta N_x \quad (29a,b)$$

For these stresses to satisfy the Tsai-Wu criterion for failure in the glass-reinforced plies, an increase in the critical traction is required of

$$\Delta N_x^* = 0.146 \; MN/m$$

corresponding to an increase in the composite stress and strain of 112.3 MPa and 1.17%, respectively, and giving final predictions for the failure strength and fracture strain of 388 MPa and 2.11%. These compare with mean experimental measurements of 381 MPa and 1.7%. If the experimentally measured compressive strengths for the carbon and glass plies are discounted and the assumption made that $X' = X$ and $Y' = Y$, it is found that the predicted hybrid failure strength and fracture strain are only very slightly modified, to 380 MPa and 2.02%, respectively. Similar calculations have been performed for the type 2a and 2b hybrid lay-ups and the results are summarised in Table 4 below.

Although very good agreement is seen between the experimentally measured and theoretically predicted hybrid failure strengths the fracture strains differ quite considerably and show opposing trends, the theoretically predicted failure strains increasing with increasing carbon content which conflicts with experimental evidence and is clearly incorrect. This is not surprising, perhaps, in view of the fact that the predicted hybrid response is based on the failure strengths of the constituent reinforcing plies rather than on their strains to failure.

A further comparison between theory and experiment may be seen in Figures 2 and 3, where the results are presented in terms of the volume fraction of carbon reinforcement, and in Figures 4, 5 and 6, where the predicted bilinear stress-strain curves are compared with the experimental data. In the case of the type 2b hybrid, Figure 6, where the discrepancy in the failure strains is most marked, the effect of modelling the damage in the failed carbon plies in terms of a reduction in the stiffness matrix by a factor of 10

TABLE 4. Comparison of Predicted and Experimentally Determined Quasi-Static Hybrid Failure Strengths and Strains.

Hybrid Lay-Up	Failure Strength (MPa)		Fracture Strain (%)	
	Experimental	Theoretical	Experimental	Theoretical
Type 1	381	388	1.7	2.11
Type 2a	413	407	1.58	2.11
Type 2b	431	419	1.45	2.19

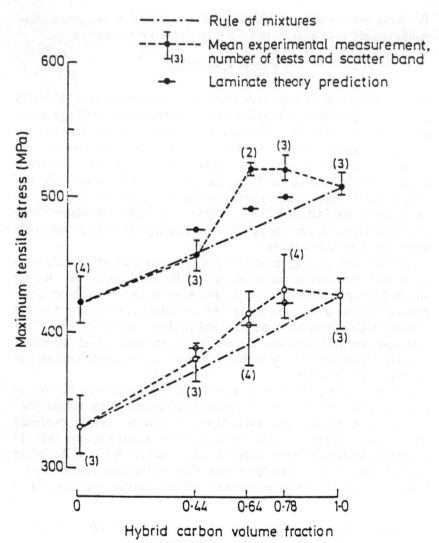

FIGURE 2. Effect of hybrid carbon fraction on maximum tensile stress comparison of theory with experiment.

----O---- quasi-static tests, loading in warp direction
----●---- impact tests, loaded in weft direction

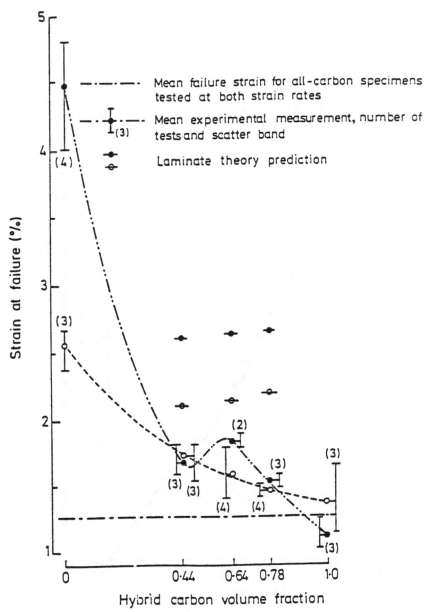

FIGURE 3. Effect of hybrid carbon fraction on strain at failure comparison of theory with experiment.

----○---- quasi-static tests, loading in warp direction
----●---- impact tests, loaded in weft direction

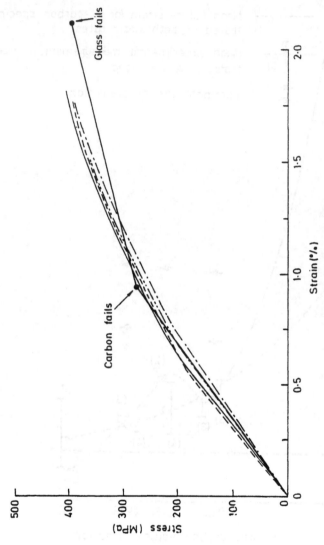

FIGURE 4. Comparison of experimentally-measured and theoretically-predicted quasi-static tensile stress-strain curves for type 1 hybrid specimens.

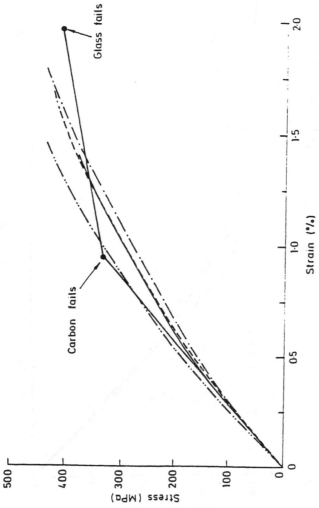

FIGURE 5. Comparison of experimentally-measured and theoretically-predicted quasi-static tensile stress-strain curves for type 2a hybrid specimens.

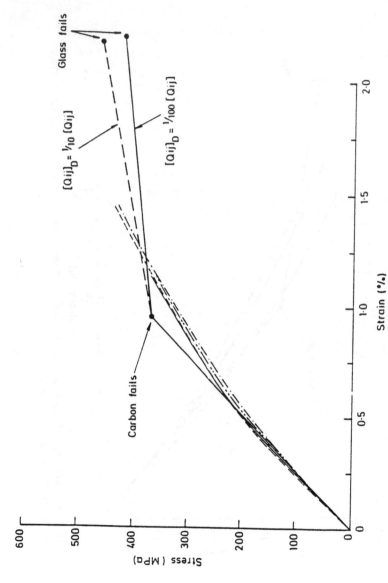

FIGURE 6. Comparison of experimentally-measured and theoretically-predicted quasi-static tensile stress-strain curves for type 2b hybrid specimens.

122

rather than 100 is shown to increase the failure strength from 419 to 459 MPa but to have no significant effect on the failure strain. It is also clear that FPF occurs at a stress well above that for the knee on the experimental stress-strain curve.

4.2 Impact Loading Rate

Similar calculations to those described above may also be performed to predict the dynamic failure strengths and fracture strains of the three hybrid lay-ups, using stiffness and strength parameters determined for the two types of reinforcing ply at the impact rate of strain. Predicted and experimentally determined failure strengths and strains for the three hybrid lay-ups are compared in Table 5 below. The predicted bilinear stress-strain response for each lay-up is compared with the corresponding experimentally determined stress-strain curves (16) in Figures 7, 8 and 9 while a comparison of the effect of the hybrid carbon fraction on the predicted and the measured tensile strengths and failure strains was shown previously in Figures 2 and 3, respectively.

5. DISCUSSION

In attempting to predict the tensile strength of the various hybrid lay-ups laminate theory is used to obtain the stress distribution in the specimen as the load is increased and failure of a given ply is taken to correspond to the applied load at which these stresses first satisfy the Tsai-Wu criterion for that ply. The strength parameters in the Tsai-Wu criterion are taken to be the tensile and compressive strengths of the two types of reinforcing ply at the given rate of loading, as determined in tests on non-hybrid carbon or glass-reinforced laminates. In practice, for tensile loading in either the warp or the weft direction, the ruling compressive stresses are very small (see Table 3), and an accurate determination of the composite compressive strength is

TABLE 5. Comparison of Predicted and Experimentally Determined Hybrid Impact Failure Strengths and Strains.

Hybrid Lay-Up	Failure Strength (MPa)		Fracture Strain (%)	
	Experimental	Theoretical	Experimental	Theoretical
Type 1	458	475	1.68	2.61
Type 2a	521	492	1.83	2.63
Type 2b	520	500	1.51	2.64

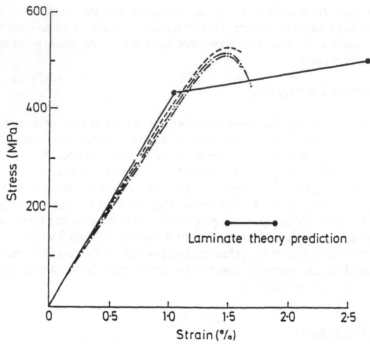

FIGURE 7. Comparison of experimentally-measured and laminate theory predicted stress-strain curves for type 2b hybrid specimens (impact loading in weft direction).

not required. However, since the Tsai-Wu formulation is based on the Hill criterion for yielding in anisotropic metals, this approach assumes an essentially linear-elastic behaviour in the individual reinforcing plies up to the point of "failure". This is not an unreasonable assumption since for both the all-carbon- and the all-glass-reinforced laminates under quasi-static loading and for the all-carbon-reinforced laminate under impact loading the experimentally determined stress-strain curves were linear almost through to the point of final fracture.

More questionable, however, are the consequent assumptions (a) that all the plies of one type "fail" simultaneously and (b) that there is a relation between the stress system at fracture (i.e., total failure) in a test on an all-carbon laminate and that at "first ply failure" in the carbon plies within the hybrid lay-up, where this is taken to mean a reduction in stiffness and a continued ability to carry load. In physical terms the initial failure of a ply is likely to correspond to a breakdown of the matrix leading to a reduced stiffness, since the axially-aligned woven tows are now able to straighten more easily. A more refined calculation might, therefore, allow this to occur under a stress system slightly below that associated with the final tensile fracture of tows. Such a calculation might also give closer agreement between the stress levels at which the experimental curves first depart from a linear-elastic response and the theoretical curves show first ply failure. However, since it is also likely that different regions within the matrix will break down and different tows within a given ply will fracture at different levels of applied stress, unless these effects can also be taken into account a more refined calculation is probably not justified.

As is clear from the preceding discussion a major limitation to the laminate theory approach for predicting the tensile strength of woven-reinforced composites is that it treats each type of reinforcing ply at the macroscopic level and so ignores the geometry of the fibre–matrix interaction and fails to identify what may be the most important properties in determining or modifying the composite impact strength, such properties as, for example, fibre/matrix interfacial bond strength or interlaminar shear strength. In particular, since it is hybrid laminates which are being modelled, the interlaminar shear strength might be expected to play an important role. In an attempt to take account of such effects a numerical technique is currently being developed to model the failure process in the hybrid laminates. Meanwhile, despite its

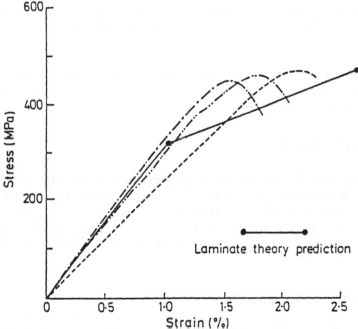

FIGURE 8. Comparison of experimentally-measured and laminate theory predicted stress-strain curves for type 1 hybrid specimens (impact loading in weft direction).

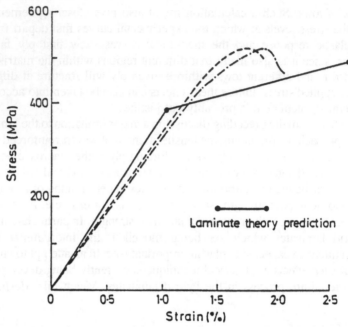

FIGURE 9. Comparison of experimentally-measured and laminate theory predicted stress-strain curves for type 2a hybrid specimens (impact loading in weft direction).

serious limitations, the laminate theory approach does seem to allow a first estimate to be made of the hybrid tensile strength at both quasi-static and impact rates of loading.

6. CONCLUSIONS

A previously developed laminate theory approach has been used to predict the tensile strength of three woven hybrid carbon/glass epoxy composites at both a quasi-static ($\sim 0.001/s$) and an impact ($\sim 1000/s$) rate of loading. The results show reasonable agreement with the experimentally determined strengths but the predicted failure strains are significant overestimates of the experimental values.

7. ACKNOWLEDGEMENT

This research was sponsored by the Air Force Office of Scientific Research, Air Force Systems Command, USAF, under Grant No. AFOSR-87-0129.

8. REFERENCES

1. Harding, J. and L. M. Welsh. 1983. "A Tensile Testing Technique for Fibre Reinforced Composites at Impact Rates of Strain", *J. Mater. Sci.*, 18:1810–1826.

2. Welsh, L. M. and J. Harding. 1985. "Dynamic Tensile Response of Unidirectionally Reinforced Carbon Epoxy and Glass Epoxy Composites", *Proc. 5th. Int. Conf. on Composite Materials*, ICCM V, TMS-AIME, pp. 1517–1531.

3. Welsh, L. M. and J. Harding. 1985. "Effect of Strain Rate on the Tensile Failure of Woven-Reinforced Polyester Resin Composites", *Proc DYMAT '85, Int. Conf. on Mechanical and Physical Behaviour of Materials under Dynamic Loading, Jour. de Physique*, Colloque C5:405–414.

4. Saka, K. and J. Harding. 1985. "The Deformation and Fracture of Hybrid Reinforced Composites under Tensile Impact", *Proc. IUTAM Colloquium on Macro- and Micro-Mechanisms of High Velocity Deformation and Fracture, Tokyo, 12–15 August*. Berlin: Springer-Verlag, pp. 97–111.

5. Xia, Y. R. and C. Ruiz. 1989. "Response to Impact of Layered Structures", *Proc. 4th. Oxford Conf. on Mechanical Properties of Materials at High Rates of Strain*, Institute of Physics.

6. Harding, J., K. Saka and M. E. C. Taylor. 1987. "Effect of Strain Rate on the Tensile Failure of Woven-Reinforced Carbon/Glass Hybrid Composites", *Proc. IMPACT '87*, (DGM Informationsgesellschafft mbH, Oberursel), 2:579–586.

7. Gol'denblat, I. I. and V. A. Kopnov. 1965. "Strength of Glass-Reinforced Plastics in the Complex Stress State", *Polymer Mechanics*, 1(2):54–59.

8. Protasov, V. D. and V. A. Kopnov. 1965. "Study of the Strength of Glass Reinforced Plastics in the Plane State of Stress", *Polymer Mechanics*, 1(5):26–28.

9. Tsai, S. W. and E. M. Wu. 1971. "A General Theory of Strength for Anisotropic Materials", *J. Comp. Materials*, 5:58–80.

10. Owen, M. J. and D. J. Rice. 1981. "Biaxial Strength Behaviour of Glass Fabric Reinforced Polyester Resins", *Composites*, 12:13–25.

11. Sendeckj, G. P. 1972. "A Brief Survey of Empirical Multiaxial Strength Criteria for Composites", *Composite Materials: Testing and Design* (Second Conference), ASTM STP 497, pp. 41–51.

12. Tsai, S. W. and H. T. Hahn. 1980. *Introduction to Composite Materials*. Lancaster, PA: Technomic Publishing Co., Inc., Chapter 7.

13. Tsai, S. W. and V. D. Azzi. 1966. "Strength of Laminated Composite Materials", *A.I.A.A. Journal*, 4:296–301.

14. Ishikawa, T. and T. W. Chou. 1982. "Stiffness and Strength Behaviour of Woven Fabric Composites", *J. of Mater. Sci.*, 17(11):3211–3220.

15. Saka, K. and J. Harding. 1986. "Behaviour of Fibre-Reinforced Composites under Dynamic Tension", Interim Report on Grant No. AFOSR-85-0218 (Oxford University Engineering Laboratory, Report No. OUEL 1654/86).

16. Saka, K. and J. Harding. 1985. "Behaviour of Fibre-Reinforced Composites under Dynamic Tension," Final Report on Grant No. AFOSR-84-0092 (Oxford University Engineering Laboratory, Report No. OUEL 1602/85).

17. Shah, S., R. K. Y. Li and J. Harding. 1988. "Modelling of the Impact Response of Fibre-Reinforced Composites", Interim Report on Grant No. AFOSR-87-0129 (Oxford University Laboratory, Report No. OUEL 1730/88).

18. Harding, J., Y. Li, K. Saka and M. E. C. Taylor. 1989. "Characterisation of the Impact Strength of Woven Carbon/Epoxy Laminates", *Proc. 4th. Oxford Conf. on Mechanical Properties of Materials at High Rates of Strain* (Institute of Physics).

19. Lempriere, B. M. 1968. "Uniaxial Testing of Orthotropic Materials", *A.I.A.A. Journal*, 6:365–368.

20. Bert, C. W., B. L. Mayberry and J. D. Ray. 1969. "Behaviour of Fibre Reinforced Plastic Laminates under Biaxial Loading", ASTM STP 460, pp. 362–380.